Springer Theses

Recognizing Outstanding Ph.D. Research

Aims and Scope

The series "Springer Theses" brings together a selection of the very best Ph.D. theses from around the world and across the physical sciences. Nominated and endorsed by two recognized specialists, each published volume has been selected for its scientific excellence and the high impact of its contents for the pertinent field of research. For greater accessibility to non-specialists, the published versions include an extended introduction, as well as a foreword by the student's supervisor explaining the special relevance of the work for the field. As a whole, the series will provide a valuable resource both for newcomers to the research fields described, and for other scientists seeking detailed background information on special questions. Finally, it provides an accredited documentation of the valuable contributions made by today's younger generation of scientists.

Theses are accepted into the series by invited nomination only and must fulfill all of the following criteria

- They must be written in good English.
- The topic should fall within the confines of Chemistry, Physics, Earth Sciences, Engineering and related interdisciplinary fields such as Materials, Nanoscience, Chemical Engineering, Complex Systems and Biophysics.
- The work reported in the thesis must represent a significant scientific advance.
- If the thesis includes previously published material, permission to reproduce this must be gained from the respective copyright holder.
- They must have been examined and passed during the 12 months prior to nomination.
- Each thesis should include a foreword by the supervisor outlining the significance of its content.
- The theses should have a clearly defined structure including an introduction accessible to scientists not expert in that particular field.

More information about this series at http://www.springer.com/series/8790

Supachai Awiphan

Exomoons to Galactic Structure

High Precision Studies with the Microlensing and Transit Methods

Doctoral Thesis accepted by
the University of Manchester, Manchester, UK

 Springer

Author
Dr. Supachai Awiphan
National Astronomical Research
  Institute of Thailand
Chiang Mai
Thailand

Supervisor
Prof. Eamonn Kerins
The University of Manchester
Manchester
UK

ISSN 2190-5053   ISSN 2190-5061 (electronic)
Springer Theses
ISBN 978-3-030-08141-6   ISBN 978-3-319-90957-8 (eBook)
https://doi.org/10.1007/978-3-319-90957-8

Printed on acid-free paper

This Springer imprint is published by the registered company Springer International Publishing AG part of Springer Nature
The registered company address is: Gewerbestrasse 11, 6330 Cham, Switzerland

"You don't have to be a mathematician to have a feel for numbers."

John Forbes Nash, Jr.
Nobel Memorial Prize laureate in Economics

Supervisor's Foreword

It is a pleasure to write this foreword for Dr. Awiphan's doctoral thesis, which provides new insights across multiple exoplanet detection techniques.

The field of exoplanets has exploded in the mere two decades since the initial discoveries. A number of techniques are now being employed to find and characterise exoplanets over a wide range of discovery space. Mapping exoplanet demographics is a vital task on the road to understanding how planets form and to place into context the architecture of our own Solar System. To this end, it will be important to bring together results from different techniques such as transits and microlensing that target hot and cool planet populations.

Dr. Awiphan's thesis spans a number of exciting strands in current exoplanet research. The thesis presents transmission spectroscopy observations in Chap. 3. The techniques of transmission spectroscopy use multi-wavelength transit observations to obtain low-resolution spectral information on exoplanet atmospheres. It is providing new insights into the structure and composition of atmospheres of hot Jupiter and, in this case, Neptunes. Using data from the 2.4 m Thai National Telescope, as well as from smaller telescopes, Dr. Awiphan presents a careful reduction of transit photometry of GJ3470b and confirms the existence of a clear Rayleigh scattering signature from its atmosphere. He also develops a transit timing variation analysis that allows him to constrain the existence of additional planets in the system.

In Chap. 4, he demonstrates through the use of detailed simulations how it will be possible to confirm the existence of potentially habitable Earth-like exomoons associated with planets within the habitable zone of the most common type of star in our Galaxy. Since moons are far more common than planets in our own Solar System, this is an important architecture to consider when considering the possibility of life. Only a few years ago this would have seemed an outlandish enterprise to undertake. But with the imminent launch of wide-area transit surveys such as TESS and PLATO, Dr. Awiphan convinces us that such goals are now within our reach.

Dr. Awiphan moves from transit timing to microlensing in Chaps. 5 and 6 by using the Besancon Galactic model as the basis for the construction of a detailed microlensing map of our Galaxy. Microlensing surveys have discovered many thousands of events, and these are enabling us to test models of the inner Galaxy in new ways. Whilst Gaia continues to map the Milky Way in unprecedented detail, it is severely limited by crowding in the inner Galaxy; this is precisely the regime in which microlensing provides greatest statistical sensitivity. Chapter 6 of Dr. Awiphan's thesis pits the most detailed microlensing model published to date with the largest completeness-corrected sample of microlensing events from the MOA survey. His analysis points to the need for further development of our model of the inner Galaxy.

Overall, the thesis is highly ambitious both in the breadth and depth of the science presented. It has been a pleasure to supervise Dr. Awiphan's work and I am confident that the work presented here will help us to fully capitalise on the next generation of transit and microlensing space-based surveys that are coming online over the next decade.

Manchester, UK   Dr. Eamonn Kerins
March 2018

Preface

Today, the search for and study of exoplanets is one of the most interesting areas of modern astronomy. Over the last two decades, the number of detected exoplanets continues to increase. As of February 2018, over 3600 exoplanets have been discovered. This thesis presents high precision studies based on the transit and microlensing methods which are used to detect hot and cool exoplanets, respectively. In Chap. 3, the transit timing variation and transmission spectroscopy observations and analyses of a hot Neptune, GJ3470b, from telescopes at Thai National Observatory, and the 0.6-metre PROMPT-8 telescope in Chile are presented, in order to investigate the possibility of a third body presence in the system and to study its atmosphere. From the transit timing variation analyses, the presence of a hot Jupiter with a period of less than 10 days or a planet with an orbital period between 2.5 and 4.0 days in GJ3470 system are excluded. From transmission spectroscopy analyses, combined optical and near-infrared transmission spectroscopy favours an H/He dominated haze (mean molecular weight 1.08 ± 0.20) with methane in the atmosphere of GJ3470b. In Chap. 4, the effects of intrinsic stellar noise to the detectability of an exomoon orbiting a transiting exoplanet are investigated using transit timing variation and transit duration variation. The effects of intrinsic stellar variation of an M-dwarf reduce the detectability correlation coefficient by 0.0–0.2 with 0.1 median reduction. With the microlensing technique, the first real-time online simulations of microlensing properties based on the Besançon Galactic model, called Manchester-Besançon Microlensing Simulator (MaBμlS), are presented in Chap. 5. We also apply it to the recent MOA-II survey results in Chap. 6. This analysis provides the best comparison of Galactic structure between a simulated Galactic model and microlensing observations. The best-fitting model between Besançon and MOA-II data provides a brown dwarf mass function slope of -0.4. The Besançon model provides only ~ 50 per cent of the measured optical depth and event rate per star at low Galactic latitude around the inner bulge. However, the revised MOA-II data are consistent with the Besançon model without any missing inner bulge population.

Chiang Mai, Thailand   Supachai Awiphan

Parts of this thesis have been published in the following articles:

- **The detectability of habitable exomoons with Kepler.**
  Awiphan, S. and Kerins, E. (2013), MNRAS, 432, 2549–2561
- **Besançon Galactic model analysis of MOA-II microlensing: evidence for a mass deficit in the inner bulge.**
  Awiphan, S., Kerins, E. and Robin, A. C. (2016), MNRAS, 456, 1666–1680
- **Transit timing variation and transmission spectroscopy analyses of the hot Neptune GJ3470b.**
  Awiphan, S., Kerins, E., Pichadee, S., Komonjinda, S., Dhillon, V. S., Rujopakarn, W., Poshyachinda, S., Marsh, T. R., Reichart, D. E., Ivarsen, K. M. and Haislip J. B. (2016), MNRAS, 463, 2574–2582

Acknowledgements

This thesis would not have been possible without the guidance and help of several people. First of all, I would like to thank my beloved country Thailand and acknowledge the Royal Thai Government Scholarship who fully sponsored my Master's and Ph.D. study at the University of Manchester. I also want to acknowledge the University of Manchester for supporting me with the President's Doctoral Scholar Award, and also my undergraduate scholarships, Development and Promotion of Science and Technology Talents Project and Junior Science Talent Project scholarships. It is my tremendous honour to receive these awards.

I would like to express my deepest gratitude and indebtedness to my supervisor, Dr. Eamonn Kerins, for his guidance, encouragement, and support throughout my Master's and Ph.D. degrees. Throughout 5 years, the skills I have learned and the numerous interesting discussions are invaluable for me.

I sincerely thank Dr. David Kipping, Dr. Annie C. Robin, Prof. Vik Dhillon, Dr. Wiphu Rujopakarn, Prof. Thomas Marsh, Prof. Keith Horne, and Dr. Andrew Markwick for their guidance in analyses and valuable suggestions with regard to this thesis.

My sincere appreciation is extended to my JBCA friends, Tianyue Chen, Yun-Huk Kim, Phrudth Jaroenjittichai, Constantinos Demetroullas, Prahesti Husnindriani, and Makiko Ban, for their kindness and sympathy, and special thanks to my co-supervisor, Dr. Tim O'Brien, my advisor, Dr. Malcolm Gray, and my first year Ph.D. progression examiner, Dr. Neal Jackson, for their support and suggestions during this study. I also wish to thank all the staff and students at JBCA who were very friendly and helpful.

I am extremely grateful to Prof. Boonrucksar Soonthornthum, ex-director of National Astronomical Research Institute of Thailand (NARIT), Dr. Saran Poshyachinda, director of NARIT, and Dr. Siramas Komonjinda and my ex-supervisor at Chiang Mai University, Thailand, who advised me to study at the University of Manchester and always support me.

I like to extend my appreciation to people from Astronomical Research Laboratory Chiang Mai University and NARIT, Dr. Suwicha Wannawichian, Dr. Utane Sawangwit, and Dr. Andrea Richichi, for their very valuable suggestion and help. Moreover, I would like to thank my colleague, Sawatkamol Pichadee, a graduated Master student at Chiang Mai University. I would also like to express my gratitude to several supportive people back home in Thailand, including Pongpichit Chuanraksasat and Napaporn A-thano who made a final review of this work.

Last but not the least, I would like to thank and share the honour of having completed this thesis with my loving mother, my brother, and my girlfriend, Nawapon Nakharutai, in gratitude for their unending support and encouragement. Without their support, this work would not have been possible. I also would like to thank everybody who was important to the success of this thesis, as well as expressing my apology that I could not mention them all personally one by one.

Finally, for two decades of education from primary school to graduate school, my school period is nearly over, but my research career just starts. Thank a sentence in 13 years ago which inspires me to these:

"Scientists in Thailand are less than 1%."

Contents

Abbreviations

AST3	Antarctic Schmidt Telescopes
AU	Astronomical Unit
CCAA	Chinese Center of Antarctic Astronomy
CCD	Charge-Coupled Device
CHEOPS	CHaracterising ExOPlanet Satellite
CoRoT	Convection, Rotation and planetary Transits
CTIO	Cerro Tololo Inter-American Observatory
DIA	Difference Image Analysis
EChO	Exoplanet CHaracterisation Observatory
EMCCD	Electron Multiplying Charge-Coupled Device
ESA	European Space Agency
ESO	European Southern Observatory
ESPRESSO	Echelle SPectrograph for Rocky Exoplanets and Stable Spectroscopic Observations
HATNet	Hungarian-made Automated Telescope Network
HST	Hubble Space Telescope
IMF	Initial Mass Function
KASI	Korea Astronomy and Space Science Institute
KELT	Kilodegree Extremely Little Telescope
KMTNet	Korea Microlensing Telescope Network
LCOGT	LasCumbres Observatory Global Telescope
MaBμlS	Manchester-Besançon Microlensing Simulator
MASCARA	Multi-site All-Sky CAmeRA
MINERVA	Miniature Exoplanet Radial Velocity Array
MMR	Mean Motion Resonance
MOA	Microlensing Observations in Astrophysics
NARIT	National Astronomical Research Institute of Thailand
NASA	National Aeronautics and Space Administration
NGTS	Next Generation Transit Survey
NISP	Near-Infrared Spectrometer and Photometer

OGLE	Optical Gravitational Lensing Experiment
PLANET	Probing Lensing Anomalies NETwork
PLATO	PLAnetary Transit and Oscillations of stars
POTS	Pre-OmegaTranS
PTV	Photocentric transit timing variation
QES	Qatar Exoplanet Survey
RMS	Root Mean Square
SAAO	South African Astronomical Observatory
SFR	Star Formation Rate
SPEARNET	Spectroscopy and Photometry of Exoplanetary Atmospheres Research NETwork
SWEEPS	Sagittarius Window Eclipsing Exoplanet Search
TDV	Transit Duration Variation
TESS	Transiting Exoplanet Survey Satellite
TIP	Transit Impact Parameter
TRAPPIST	TRAnsiting Planets and PlanetesImals Small Telescope
TrES	Transatlantic Exoplanet Survey
TTV	Transit Timing Variation
VIS	VISible and near-infrared imaging channels instrument
WASP	Wide Angle Search for Planets
WFIRST	Wide-Field Infrared Survey Telescope

List of Figures

List of Tables

Chapter 1
Introduction

Over the last two decades, the search for and study of exoplanets, planets outside the Solar System, has been one of the most dynamic research fields of modern astronomy. This is not only because of the potential that living things may exist on them, but also to further our understanding of the origin of planetary systems. As of February 2018, more than 3600 planets have been confirmed from ground-based and space-based observation,[1] including some super-Earth planets, planets with masses between 1 and 10 $M_\oplus$, which have potential to support life.

The number of detected exoplanets increased dramatically in recent years due to the success of the NASA *Kepler* mission. The *Kepler* space telescope was launched in 2009 in order to discover transiting exoplanets in or near the habitable zone and to determine the distribution of sizes and shapes of the orbits of the planets. By 2016, more than 2300 exoplanets and over 4500 planet candidates have been discovered by the *Kepler* mission based on the entire 48 month *Kepler* dataset (Coughlin et al. 2016; Morton et al. 2016). A large majority of the candidates are expected to be planets (Fressin et al. 2013).

The first discovery of exoplanets was in 1992 when the first confirmed exoplanet was found by Wolszczan and Frail (1992). They announced the discovery of planets around a pulsar, PSR 1257 + 12, by using the pulsar timing method. In 1995, the first exoplanet orbiting an ordinary star 51 Peg b, a 0.5 Jupiter-mass exoplanet with 4.2 day orbital period, was detected using the radial velocity method, which looks for the periodic Doppler shifts in stellar spectra (Fig. 1.1) (Mayor and Queloz 1995). Several year later, the first transiting exoplanet, HD 209458b, was discovered (Charbonneau et al. 2000). The transit technique is a detection method which looks for periodic dimming in stellar brightness (Fig. 1.2). Following on from the discoveries in the 1990s, several different methods have been developed to discover exoplanets, including gravitational microlensing and direct imaging. To date, the majority of the detected exoplanets have been found using the transit technique. However, the

---

[1] See https://exoplanetarchive.ipac.caltech.edu/.

© Springer International Publishing AG, part of Springer Nature 2018
S. Awiphan, *Exomoons to Galactic Structure*, Springer Theses,
https://doi.org/10.1007/978-3-319-90957-8_1

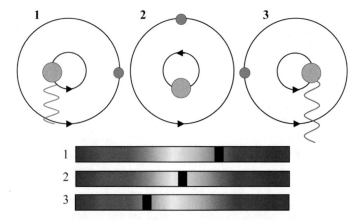

Fig. 1.1 Schematic showing the radial velocity detection method. The star is indicated in orange and the planet in blue. The star orbits around the center of mass due to the gravitational interaction of the planet. The Doppler shift of the starlight causes stellar spectral lines (black line) to move periodically red ward or blue ward

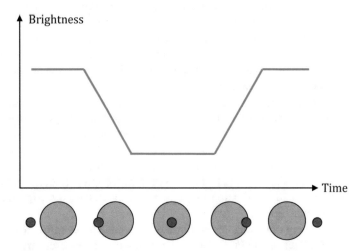

Fig. 1.2 Schematic showing transit detection method. The star is indicated in orange and the planet in red. When the planet passes the star in the direction of observer line-of-sight, the planet blocks the star's light and causes a dip in stellar brightness

transit method can only obtain some of the planetary parameters, such as the planet-star radius ratio and orbital inclination. In order to obtain all planetary parameters, such as exact mass, radius and mean density, additional data from radial velocity and other methods are often used (Udry and Santos 2007).

Although more than half of the detected exoplanets have been discovered by the transit method, there are hot Jupiters, Jupiter-type exoplanets orbiting close to their host stars (≤ 0.1 AU) (Udry and Santos 2007). Hot Jupiters dominate the statistics

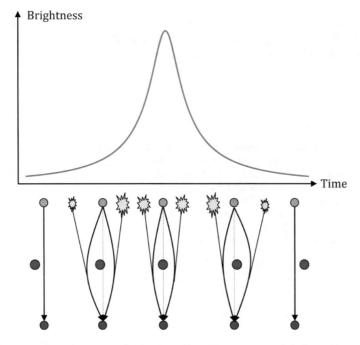

Fig. 1.3 Schematic showing the microlensing effect. The source star is indicated in orange and the lens object in red. The lens moves into the observer's line-of-sight towards the source. The brightness of the source is magnified due to the gravity of the lens

because the method is more sensitive to short orbital period exoplanets with relatively large planet-star radius ratio. Although the transiting hot Jupiters provide information on planets in extreme environments, the vast majority of exoplanets are of lower mass with longer orbital period.

The microlensing technique has proven to be efficient for discovering long-period Earth-mass exoplanets (Gould et al. 2010). Microlensing is a special case of gravitational lensing which can be detected by the change in brightness of a background star as a foreground planetary host passes by (Fig. 1.3). The stellar brightness is magnified due to the gravity of the foreground lens. If the lens is a planet orbiting around a star, the magnified light will be perturbed. This method can also be used to detect free-floating planets.

In this thesis, we present simulations and observations which showcase the transit and microlensing methods. The structure of the thesis is as follows. In Chap. 2, a historical, basic theory background of high time resolution exoplanet detection techniques, such as radial velocity, transit and microlensing is provided. In Chap. 3, transmission spectroscopy data of the hot-Neptune GJ3470b are presented and analysed in Chap. 3. Effects of intrinsic stellar noise on the detectability of exomoons with *Kepler*-class photometry are investigated in Chap. 4. An online real-time microlensing simulation, based on the Besançon Galactic model is presented in Chap. 5.

The Manchester-Besançon Microlensing Simulator (MaBμlS) is calibrated to the optical depth, average time scale and event rate, from the MOA-II survey in Chap. 6. Finally, my future works are discussed in Chap. 7.

References

Charbonneau D, Brown TM, Latham DW, Mayor M (2000) ApJ 529:L45
Coughlin JL, Mullally F, Thompson SE et al (2016) ApJS 224:12
Fressin F, Torres G, Charbonneau D et al (2013) ApJ 766:81
Gould A, Dong S, Gaudi BS et al (2010) ApJ 720:1073
Mayor M, Queloz D (1995) Nature 378:355
Morton TD, Bryson ST, Coughlin JL et al (2016) ApJ 822:86
Udry S, Santos NC (2007) ARA&A 45:397
Wolszczan A, Frail DA (1992) Nature 355:145

Chapter 2
Basic Theory Exoplanet Detection

2.1 Exoplanet Detection Methods

During the last two decades, more than 3,600 exoplanets have been found and the number is still rapidly increasing. The number of detections is very small compared to the number of stars in the Universe, although every stars may host at least one planet (Cassan et al. 2012). Exoplanets are very difficult to detect, due to their extremely low light emission compared to their host stars and the fact that the separation between the star and planet on the sky is very small. In order to detect exoplanets, a number of techniques have been developed, including radial velocity, transit, microlensing, pulsar timing, direct imaging and astrometry methods. Each method has its own advantages and disadvantages.

2.1.1 Radial Velocity Method

Radial velocity is a technique which detects the star wobble around the system centre-of-mass due to the gravitational interaction of nearby planets. The motion of the star produces the change in velocity along the line of sight to the star that causes a periodic shift of the lines in the star's spectrum. This effect can be measured accurately by observing with a high-resolution spectrograph.

This method favours massive planets which have a short orbital period, because of the higher amplitude of the radial velocity signal, K_*. The amplitude of the radial velocity signal of a star of mass M_* with planet of mass M_p orbiting around with period P_p, is defined by,

$$K_* = \frac{M_p \sin i_p}{M_*} \frac{2\pi a_p}{P_p} , \qquad (2.1)$$

where a_p is the planet-host separation and i_p is the orbital inclination. The radial velocity technique has a disadvantage of not being able to determine the true mass of planets, but only their minimum mass, $M_p \sin i_p$.

© Springer International Publishing AG, part of Springer Nature 2018
S. Awiphan, *Exomoons to Galactic Structure*, Springer Theses,
https://doi.org/10.1007/978-3-319-90957-8_2

However, this method has proved to be a successful exoplanet detection technique, because a number of exoplanets discovered from the ground have been detected by this method. Since the first detection of an exoplanet via radial velocity in 1995, 51 Peg b (Fig. 2.1, Mayor and Queloz (1995)), more than 700 planets have been discovered by using this method.

Nowadays, with improvements in the sensitivity of the technique and instruments, a high-resolution spectrograph, such as ESPRESSO (Echelle SPectrograph for Rocky Exoplanets and Stable Spectroscopic Observations), can reach a precision of ~ 0.1 m.s^{-1} allowing Earth-like planets in the habitable zones to be detected (Pepe et al. 2014).

2.1.2   Transit Method

The transit technique relies on the detection of periodic dips in the stellar light curve. When a planet passes in front of its host star, the star flux temporarily decreases due to blocking by the planet. To date, the measurements of stellar brightness have a precision better than 0.1% for ground-based telescopes. For space-based telescopes, such as *Kepler*, the precision is 20 parts per million relative precision in 6.5 h for a 12th magnitude G-type main-sequence star, sufficient to detect an Earth-mass planet transiting a solar-type star (Koch et al. 2010; Jenkins et al. 2015).

Fig. 2.1 Original data of 51 Peg shows variation in orbital motion due to 51 Peg b's gravitational interaction (Mayor and Queloz 1995)

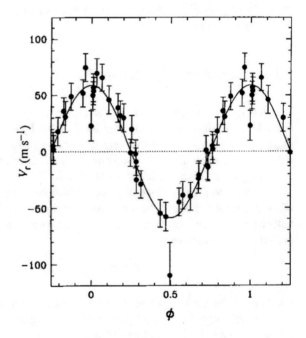

The transit technique provides information about the system by fitting a physical model to the light curve. The planetary parameters, such as planet's orbital inclination and radius, can be determined when combined with an estimate of the host star radius. Combining transit observations with radial velocity measurements, the true planetary mass and density can be calculated. Therefore, the follow-up radial velocity observations are important for determining transiting exoplanet parameters.

However, the transit technique has the problem of a large number of false-positive detections caused by grazing eclipsing binaries, low-mass stellar objects or blended stellar systems (Díaz et al. 2011). Moreover, the transit technique can only detect planets that orbit in front of their host stars. The detection probability is about 10% for hot Jupiters and 0.5% for Earth-like planets (See Sect. 3.1.1). The number of detections is also biased toward short orbital period exoplanets (≤ 10 days) and exoplanets with large radii.

The first detected transiting exoplanet, HD 209458b, was discovered by Charbonneau et al. (2000) and Henry et al. (2000), although HD 209458b was discovered using the radial velocity method. Its transit was observed using a 0.8 m telescope by Henry et al. (2000) and a 1.0 m telescope by Charbonneau et al. (2000). After the first detection, a large proportion of exoplanet detections have been discovered using the transit method through ground-based transit searches, such as OGLE (Udalski 2009), WASP (Pollacco et al. 2006; Smith 2014) and HAT (Bakos et al. 2011, 2013). Nevertheless, nowadays, major transit discoveries are expected from space, such as the *Kepler* mission, which has discovered more than 2,300 confirmed exoplanets and 4,500 planetary candidates (Coughlin et al. 2016; Morton et al. 2016).

2.1.3 Gravitational Microlensing Method

Gravitational microlensing detection of exoplanets was proposed by Mao and Paczynski (1991) and Gould and Loeb (1992). It is the only known method capable of discovering planets at truly great distances from the Earth. Microlensing can find planets orbiting stars near the centre of the Galaxy, thousands of light-years away, whereas radial velocity and transit methods can detect planets only in our Galactic neighbourhood. It is also the only method which can detect free-floating planets. Microlensing is most sensitive to cold planets, in the outer regions of systems beyond the snow line, the distance from the star where solid ice grains condense from hydrogen compounds (Gaudi 2012; Kennedy and Kenyon 2008; Lecar et al. 2006; Lin 2008).

Microlensing is based on the gravitational lens effect which occurs when a lensing star moves in front of the source star. The light paths of the source are bent by the lensing star, which can magnify and demagnify the image of the source star. If the lensing star has a planet orbiting around it, the planet can perturb the light and lead to additional spikes in the light curve (Mao and Paczynski 1991).

In 2004, the first planet which was discovered using microlensing was OGLE 2003-BLG-235/MOA 2003-BLG-53 (Bond et al. 2004). Currently, more than 60 planets have been detected using this method, including a potential sub-Earth-mass moon (Bennett et al. 2014). However, the number of planet detections using microlensing will increase in the future due to the ongoing main microlensing surveys, MOA (Bond et al. 2002), OGLE (Udalski et al. 2008) and KMTNet (Henderson et al. 2014).

2.1.4   Pulsar Timing Method

The first detection of exoplanets, PSR 1257 + 12B and PSR 1257 + 12C was made in 1992 by using pulsar timing (Wolszczan and Frail 1992) (Fig. 2.2). Although this method was not originally designed for the detection of planets, pulsar timing has a capability to detect smaller planets down to less than Earth's mass. It uses the changing position of the pulsar, as a result of gravitational interaction, which causes a periodic change in the distance and light-travel time between the pulsar and observer. However, since planets orbiting a pulsar are exotic objects and are presumably very rare, the chance of finding large numbers of planets with it seems small. In this case, the planet might be formed from the stellar explosion and were not present during the stellar phase.

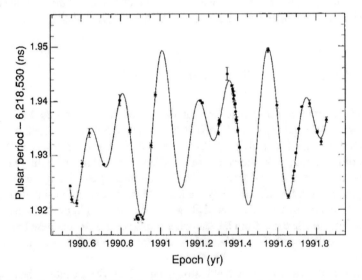

Fig. 2.2 Data of pulsar, PSR 1257 + 12, shows the variation in period due to the interaction of the first two detected exoplanets (Wolszczan and Frail 1992)

2.1.5 *Direct Imaging Method*

Direct imaging is one of the hardest exoplanet detection techniques. It is challenging for two main reasons. First, the planets are extremely faint light sources compared to their host stars. Second, the planets are generally located very close to their parent stars. Currently, special techniques such as coronagraphy and adaptive optics are used to implement this method. They are sensitive to distant hot planets around young, nearby stars which are uncommon (Masciadri and Raga 2004; Nielsen et al. 2008). The advantage of direct imaging is that it provides interesting information about the planet, including the planet's exact orbital parameters and surface properties.

Chauvin et al. (2004) discovered the first directly imaged exoplanet, 2M1207b, orbiting brown dwarf, 2MASSWJ 1207334-393254. The planet has a mass of 5 Jupiter mass with temperature in the range 1100–1300K (Chauvin et al. 2005). Currently, this technique has detected more than 40 planets, including a multi-planet system, HR 8799 (Marois et al. 2008).

2.1.6 *Astrometry Method*

Astrometry uses the idea that the gravitational interaction between star and planet causes the star and planet to orbit around planet-star barycentre. It measures a periodic variation in the position of the star on the plane of the sky, subtracting out the star's apparent motion due to the yearly parallax motion and the projection of its real proper motion through space. The main advantage of the astrometry method is that it provides an accurate estimate of a planet's mass. However, it is most effective when the orbital plane is "face on", or perpendicular to the observer's line of sight.

Strand (1943) and Reuyl and Holmberg (1943) claimed the first detected planets orbiting the stars, 61 Cyg and 70 Oph, by using the astrometric technique. However, the recent ground-based astrometry observation results show that the evidence for planets around 61 Cyg and 70 Oph has been proved incorrect (Heintz 1978). Nowadays, there are some exoplanets that have their mass determined via the astrometric technique: GI 876b (Benedict et al. 2002) HD 136118b (Martioli et al. 2010) and HD 38529c (Benedict et al. 2010). However, the first exoplanet successfully detected astrometrically came in 2010 (Pravdo and Shaklan 2009; Bean et al. 2010c).

2.2 Exoplanet Detection Programs

Since the discovery of the first exoplanet, there are various detection programs running in order to detect exoplanets. In this thesis, we focus on the transit and microlensing exoplanet detection methods. Both methods have many past, on-going and future programs running, both ground-based and space-based. The description of major programs are listed in this section (Table 2.1).

2.2.1 Ground-Based Transit Surveys

WASP

The WASP (Wide Angle Search for Planets) is the United Kingdom's leading exoplanet detection program via the transit technique. The two WASP observatories each consist of eight 200 mm camera lens robotic telescopes (0.11 m aperture) located in the northern and southern hemispheres. Each telescope has a field of view 7.8 × 7.8°. SuperWASP-North is located at the Observatorio del Roque de los Muchacho, La Palma, whilst SuperWASP-South is located at the South African Astronomical Observatory (SAAO), Sutherland, South Africa. The camera comprises Andor e2v 2K × 2K CCDs on a single equatorial fork mount. Each night, the two SuperWASP telescopes survey around a dozen fields per night at similar declination covering one

Table 2.1 Discoveries of exoplanets by transit and microlensing methods (As of February 2018)

Experiment	Method	Type	Number
OGLE	Transit	Ground-based	8
WASP	Transit	Ground-based	141*
HAT	Transit	Ground-based	62*
HATS	Transit	Ground-based	43
TrES	Transit	Ground-based	5
XO	Transit	Ground-based	6
KELT	Transit	Ground-based	19
QES	Transit	Ground-based	6
TRAPPIST	Transit	Ground-based	7
MASCARA	Transit	Ground-based	1
NGTS	Transit	Ground-based	1
POTS	Transit	Ground-based	1
Kepler	Transit	Space-based	2341
K2	Transit	Space-based	197
CoRoT	Transit	Space-based	34
SWEEPS	Transit	Space-based	2
OGLE	Microlensing	Ground-based	42†
MOA	Microlensing	Ground-based	24†
KMTNet	Microlensing	Ground-based	1

*: 3 transiting exoplanets are jointly discovered by WASP and HAT
†: 4 microlensing exoplanets are jointly discovered by OGLE and MOA

hour in right ascension, except the crowded Galactic plane (Pollacco et al. 2006; Smith 2014). Two half-minute exposures are taken at each visit with around 10 min cadence for a full season (~5 months).

In 2010, the first WASP public data was released. These data contains light curves, raw and calibrated images of the first three observing seasons (2004–2008) containing more than 30 detected exoplanets (Butters et al. 2010). To date, WASP has discovered more than 140 transiting exoplanets which is the largest number of transiting exoplanets from the ground. Highlights include WASP-17b, the first retrograde exoplanet (Anderson et al. 2010; Triaud et al. 2010) and WASP-33b, the first planet orbiting an A-type star (Collier Cameron et al. 2010). Nowadays, SuperWASP-North observes in fewer fields with higher cadence, in order to find smaller exoplanets. SuperWASP-South currently observes with new wider angle lenses (85 mm lenses) with larger field of view (18 × 18°) to discover bright transiting exoplanets (Smith 2014).

HATNet

The HATNet (Hungarian-made Automated Telescope Network) was developed in 1991 by Hungarian astronomers. The network of 6 small fully automated telescopes (4 at Fred Lawrence Whipple Observatory, Arizona, and 2 at Mauna Kea, Hawaii) with 0.11 m lenses was established in 2003 (Bakos et al. 2004, 2011). The 4K × 4K CCDs with Sloan r filters were used. HATNet monitors 838 non-overlapping fields. Fields are chosen for observation for 3 months based on visibility and various factors (Bakos et al. 2011). HATNet discovered its first transiting exoplanet in 2006, HAT-P-1b (Bakos et al. 2007). To date, HATNet has discovered about 60 confirmed transiting exoplanets, including HAT-P-32b and HAT-P-33b, which are super inflated hot Jupiters (Hartman et al. 2011).

HATSouth is the world's first telescope network which can monitor year-round 24 h. The network of six robotic telescopes with four 0.18 m lenses started observation in 2009. Each lens covers a field of view 4.18 × 4.18° of (8.2 × 8.2° in total) The telescopes are located on three sites: Las Campanas Observatory in Chile, the High Energy Stereoscopic System in Namibia and Siding Springs Observatory in Australia (Bakos et al. 2009, 2013). The HATSouth locations enable 24 h observations which gives HATSouth an order of magnitude higher sensitivity than HATNet to planets with period longer than 10 days. HATSouth monitors 12 high priority fields each year (2 months per field). To date, more than 40 transiting exoplanets have been discovered by HATSouth.

TrES

Trans-Atlantic Exoplanet Survey (TrES) was the first wide-field transit survey. TrES compressed three 0.1 m wide-field (6° and a plate scale of 11") telescopes at Lowell Observatory, Arizona; Palomar Observatory; California and Tenerife in the Canary Islands (Alonso et al. 2004). They usually observed the same field with 2 min exposure continuously in red filters for 2 months. After the observation, the data was binned to 9 min resolution. TrES discovered the second transiting exoplanet, TrES-1b: a 0.75 Jupiter-mass hot Jupiter (Alonso et al. 2004). The survey ceased operation in 2011 having discovered five hot Jupiter exoplanets in total (Mandushev et al. 2011).

XO

XO project was designed to find hot Jupiters transiting bright stars ($9 \leq V \leq 12$) with a precision of ~ 10 mmag per measurement (McCullough et al. 2005). In 2003, the survey started with two 0.1-m diameter telescope with 4K × 4K CCDs at Haleakala summit on Maui, Hawaii. The telescopes simultaneously observed the same field with a $7.2 \times 7.2^{\circ 2}$ field-of-view. Each star was observed every 10 min for 4 months. The first exoplanet discovered with XO is XO-1b, a hot Jupiter around G1 V star (McCullough et al. 2006). The project discovered 6 exoplanets in total. The latest XO exoplanet is XO-6b, a Jupiter-sized exoplanet (Crouzet et al. 2017).

KELT

Kilo-degree Extremely Little Telescope (KELT) is a transiting exoplanet survey of bright host stars ($8 \leq V \leq 10$) (Pepper et al. 2003, 2007). KELT-North uses a single very wide field ($26 \times 26^{\circ 2}$) 42 mm aperture (focal length f/1.9) telescope at Winer Observatory, Sonoita, Arizona. The survey started operation in 2005. The telescope covers 13 survey fields at declination of $+30^{\circ}$ (about 25% of the Northern hemisphere). KELT-South telescope, a duplicate telescope of KELT-North for the Southern sky, is located at the South African Astronomical Observatory, Sutherland (Pepper et al. 2012). KELT-South started operations in 2012. The first transiting exoplanet discovered by KELT is KELT-1b, a highly inflated hot Jupiter (Siverd et al. 2012). To date, more than 15 transit exoplanets were discovered by KELT, including KELT-10b, the first transiting exoplanet from the KELT-South telescope (Kuhn et al. 2016).

QES

The Qatar Exoplanet Survey (QES) is a project to discover hot Jupiters and hot Neptunes transiting bright stars (Alsubai et al. 2013). QES targets bright host stars (magnitude range from 8–15th) in order to fill the brightness gap between OGLE and the other wide-field surveys. The first QES site is in New Mexico. The project consists of four 400 mm (focal length f/2.8) and one 200 mm (focal length f/2.0) cameras with 4K × 4K CCDs. The lenses cover $5.3 \times 5.3^{\circ 2}$ and $11 \times 11^{\circ 2}$ field-of view for the 400 and 200 mm cameras, respectively. The cameras take 100 s (400 mm cameras) and 60 s (200 mm camera) exposures with an RMS accuracy of 1%. The data reduction uses the difference image analysis (DIA) technique. Since the start of operation in 2009, QES discovered six transiting exoplanets.

TRAPPIST

TRAnsiting Planets and PlanetesImals Small Telescope (TRAPPIST) is a pair Belgian 60 cm robotic telescopes for transiting exoplanet survey. TRAPPIST-South telescope located at La Silla Observatory, Chile and TRAPPIST-North located at Oukaimden Observatory, Morocco. The telescopes attached with 2K × 2K CCD camera with a field of view of 22×22 arcmin (Gillon et al. 2011, 2013). Most of the observations are obtained through an $I + z$ filter. In 2016, TRAPPIST discovered three transiting exoplanets around an ultra-cool red dwarf, TRAPPIST-1b,

TRAPPIST-1c and TRAPPIST-1d (Gillon et al. 2016). On 22 February 2017, the team announced that TRAPPIST-1 is composed of seven temperate terrestrial planets. Three planets: TRAPPIST-1e, TRAPPIST-1f and TRAPPIST-1g, are located in the habitable zone which set a new record for greatest number of habitable-zone planets found around a single star outside our solar system (Gillon et al. 2017).

MASCARA

The Multi-site All-Sky CAmeRA (MASCARA) is a ground-based transit survey that targets bright host stars with magnitudes $4 < m_v < 8.4$ (Talens et al. 2017b). MAS-CARA consists of two stations in the northern hemisphere at La Palma, Canary Islands, Spain, and in the southern hemisphere at La Silla Observatory, Chile. Each station consists of five cameras to cover entire local sky. In 2017, MASCARA discovered a hot Jupiter transiting a bright A-type star, MASCARA-1b (Talens et al. 2017a).

NGTS

Next Generation Transit Survey (NGTS) is a new ground-based sky survey designed to discover transiting Neptune and super-Earth orbiting nearby bright stars ($V < 13$) (Wheatley et al. 2013). The survey consists of twelve 20 cm f/2.8 telescopes with red sensitive CCDs at ESO Paranal Observatory, Chile. NGTS began the operation in early 2015. The survey covers more than 16 times of *Kepler* field (96 deg², 8 deg² each). The first exoplanet discovered with NGTS is NGTS-1b, a hot Jupiter transiting an M-dwarf (Bayliss et al. 2017).

POTS

Pre-OmegaTranS (POTS) project was a ground-based transit survey with ESO Wide Field Imager at the 2.2 m telescope at the La Silla observatory in the year 2006–2008. The CCD covered field-of-view are 34×34 arcmin and the images were centred at RA $= 13^h35^m41.^s6$ and Dec $= -66°42'21''$. The observations were obtained in the U, B, V, R, and I bands. POTS project discovered a hot Jupiter transiting a mid-K dwarf, POTS-1b (Koppenhoefer et al. 2013).

MINERVA

Miniature Exoplanet Radial Velocity Array (MINERVA) is a project to discover and characterize exoplanets around nearby bright stars with both high-resolution spectroscopy and photometry (Swift et al. 2015). MINERVA consists of an array of four 0.7 m telescopes at the Fred Lawrence Whipple Observatory, Mt. Hopkins, Arizona. In 2015, photometric observation was begun with $2K \times 2K$ CCDs.

LCOGT

LasCumbres Observatory Global Telescope (LCOGT) is a set of robotic telescopes around the globe dedicated to time-domain astronomy, such as: supernovae, exoplanets and pulsating stars (Shporer et al. 2011; Brown 2013). LCOGT network consists of two 2.0 m telescopes at Mt. Haleakala, Maui, Hawaii and Siding Spring Observatory, Australia, and nine 1.0 m telescopes: 3 at Cerro Tololo Inter-American Observatory, Chile, 3 at South African Astronomical Observatory, Sutherland, South

Africa, 2 at Siding Springs Observatory, Australia and 1 at McDonald Observatory, Fort Davis, Texas, USA. LCOGT collaborates with other transiting exoplanet surveys to observe exoplanet candidates and also monitors microlensing events for RoboNet (Tsapras et al. 2009).

2.2.2 Space-Based Transit Missions

CoRoT

The CoRoT (Convection, Rotation and planetary Transits) is a space-based project focused on asteroseismology of variable stars. However, after the discovery of the first hot Jupiter, CoRoT now includes exoplanet detection in the programme. The 27-cm telescope with a $2.8 \times 2.8°^2$ camera was launched in December 2006. The telescope has 4 CCDs, two are for the exoplanet program and two are for the seismology program, though one detector failed in 2009.

CoRoT can detect planets around stars in a magnitude range between 12 and 16 which include super-Earths orbiting stars $V < 14$. CoRoT typically observed 3–4 different fields per year for an average of 78 continuous days each. CoRoT detected its first transiting exoplanet in 2007 (Barge et al. 2008). CoRoT has detected more than 19 exoplanets (Moutou and Deleuil 2015), including the first transiting super-Earth, CoRoT-7b (Léger et al. 2009). However, the mission ended in 2012 when a spacecraft data processing unit failed.

Kepler

Kepler is a NASA space mission designed to discover and determine the frequency of exoplanets, their characteristics and their host star characteristics using the transit method. It is a 0.95-m telescope with 42 CCDs (field of view of 105 deg^2) and was launched in March 2009. *Kepler* detected transiting exoplanets by monitoring the brightness of around 150,000 stars in the same field of view (in the Cygnus-Lyra region) in a wavelength band between 450 and 900 nm simultaneously for 3.5 years. The minimum magnitude limit is $V = 14$. Most of the stars were observed in a long cadence mode (29.4 min), while some stars were observed in short cadence (58.8 s) with 6 s readout time (Borucki et al. 2008).

In February 2011, the *Kepler* mission released data for 156,453 stars, including 1,235 planetary candidates in 997 systems. 68 candidates are Earth-sized planets ($Rp < 1.25R_{\oplus}$) and 288 planets are super-Earth size ($1.25R_{\oplus} \leq Rp < 2R_{\oplus}$) (Borucki et al. 2011). In early 2012, more than 1,000 additional candidates were announced (Batalha et al. 2013). In May 2014, the *Kepler* nominal mission ceased operation due to failure of second reaction wheel which made it impossible to point accurately. Based on the entire 48 month *Kepler* dataset, over 2,300 confirmed exoplanets and 4,500 planet candidates have been discovered by the *Kepler* mission (Coughlin et al. 2016; Morton et al. 2016).

However, *Kepler* was reused with only two functioning reaction wheels on the mission called *K2*. The *K2* mission uses the solar flux to help to stabilize the spacecraft, therefore the telescope points in the ecliptic plane in *K2* mission (Howell et al. 2014). The *K2* mission began in June 2014 and more than 10 fields campaigns are set for observation for about 80 days in each. During the *K2* mission, more than 190 transiting exoplanets were discovered.

For microlensing, *K2* Campaign 9 conducted a survey toward the Galactic bulge between 7th April and 1st July 2016 (Henderson et al. 2016). During the campaign, more than 120 microlensing events, including microlensing exoplanets, are expected to be detected.

SWEEPS

Sagittarius Window Eclipsing Exoplanet Search (SWEEPS) was a survey with the Hubble Space Telescope between 22–29 February 2014 (Sahu et al. 2006). The Advanced Camera for Surveys - Wide Field Channel with V (F606W) and I (814W) filters was used to observe Sagittarius Windows. 16 planetary candidates with orbital period between 0.4 and 4.2 days, including two confirmed exoplanets, SWEEPS-4b and SWEEP-11b, were discovered.

TESS

Transiting Exoplanet Survey Satellite (TESS) is a NASA space telescope designed to search for exoplanets using the transit technique (Ricker et al. 2014). The mission aim will be to detect exoplanets orbiting nearby bright stars for two years. TESS will be launched in April 2018 and will monitor more than 200,000 main-sequence stars over the full sky. Each star will be observed for an interval from one month (the ecliptic plane) to one year (the ecliptic poles), depending on the stellar ecliptic latitude. The full frame images will be recorded every 30 min. TESS consists of four wide-field ($24 \times 24^{\circ 2}$) 100 mm lenses with $4K \times 4K$ CCDs. It will observe within a 600 to 1000 nm bandpass.

PLATO

The PLATO (PLAnetary Transit and Oscillations of stars), an ESA Medium class mission, has been selected for the M3 launch slot for the Cosmic Visions 2015–2025 programme (Rauer et al. 2014). It has a primary goal to discover and characterize exoplanets with high precision using 34 wide-field telescopes (32 with 25 s cadence and 2 with 2.5 s cadence) which will cover a field-of-view of 2232 deg^2. PLATO will observe stars in a magnitude range between 4 and 16. It will detect planets down to Earth-sized orbiting F to M main sequence stars in their habitable zone. Moreover, it will determine the radius and mass of the host stars and the planets with an accuracy of 1% and the age of the systems with accuracy better than 10%, giving the potential to detect exomoons and planet rings.

CHEOPS

CHaracterising ExOPlanet Satellite (CHEOPS) is the first S-class ESA mission which was selected in October 2012 and will be launched in 2018 (Broeg et al. 2013). CHEOPS is a small photometric observatory (32 cm telescope) on low Earth orbit.

It will perform ultra-high precision photometry of bright stars (150 ppm/min for a 9th magnitude star) in order to detect Earth-sized transiting exoplanet and also determine accurate radii for planets, which have mass measurements from ground-based spectroscopy surveys. It will observe 500 targets during a 3.5 year mission.

2.2.3 Ground-Based Microlensing Surveys

OGLE

OGLE (Optical Gravitational Lensing Experiment) is a survey originally designed for discovering dark matter using the mircrolensing technique, but the main focus now is to discover microlensing and transiting exoplanets. The main targets of this project are the Magellanic Clouds and the Galactic Bulge, due to the large number of stars that can be used for microlensing. The project started in 1992 and three phases of the project have been completed: OGLE-I (1992–1995), OGLE-II (1996–2000), and OGLE-III (2001–2009). The fourth is still running.

In the project's first phase (OGLE-I), the 1-m Swope telescope at the Observatory of Las Campanas, Chile was used. The first microlensing event by a binary object was discovered in this phase (Udalski et al. 1994). In second phase, the 1.3 m Warsaw telescope at Las Campanas Observatory, Chile, was used to detect microlensing events, with main targets the Large Magellanic Cloud, the Small Magellanic Cloud, the Galactic Bulge and the Galactic Disk (Udalski et al. 1997). The data from the observations of OGLE in this phase was used to search for exoplanets by follow-up projects such as the PLANET project (Dominik et al. 2002). The six regular campaign observations with eight-chip mosaic CCD camera within four 35' × 35' fields started in third phase (Udalski et al. 2002; Udalski 2009), where the first microlensing exoplanet was detected (Bond et al. 2004). Since 2010, OGLE-IV started with new a 32 chip mosaic camera which increased the observing capabilities (Udalski 2009).

In 2003, with higher cadence observation in the same fields, the first transiting exoplanet discovered by OGLE, OGLE-TR-56b, was announced (Sasselov 2003). To date, 8 transiting exoplanets have been discovered by OGLE. From 2001–2009, OGLE detected 3,718 unique microlensing events (Wyrzykowski et al. 2015), which include more than 40 confirmed microlensing exoplanets. In 2017, OGLE-IV team published data of 2,617 microlensing events during 2010–2015 observational season and found that there is no large population of unbound or wide-orbit Jupiter-mass exoplanets (Mróz et al. 2017).

MOA

MOA (Microlensing Observations in Astrophysics) is a collaborative project between researchers in New Zealand and Japan which started in 1995, using the 0.61 m telescope at Mt. John observatory in New Zealand (MOA-I) (Sumi 2010). This telescope was replaced by a 1.8 m telescope (MOA-II) equipped with a 8K × 10K

CCD camera with a 2.18 deg^2 field of view. The observation cadence is 15–45 min depending on the observed field. This project uses the microlensing technique to observe dark matter, exoplanets and stellar atmospheres. Images are reduced by difference image analysis (DIA) in real time. The first microlensing exoplanet, OGLE-2003-BLG-235/MOA-2003-BLG-53m, a 1.5 Jupiter-mass planet, was discovered by MOA survey with OGLE collaboration (Bond et al. 2004). In 2013, MOA-II team published data of 474 high quality microlensing events during 2006–2007 observational season (Sumi et al. 2013). At present, more than 20 microlensing exoplanets, including possible free floating planets (Sumi et al. 2011) and a potential sub-Earth-mass moon orbiting a 4 Jupiter-mass free-floating exoplanet (Bennett et al. 2014), have been discovered by MOA.

KMTNet

KMTNet (Korea Microlensing Telescope Network) is a project of KASI (Korea Astronomy and Space Science Institute), which consists of three 1.6 m wide-field (4 deg^2) telescopes in Cerro Tololo Inter-American Observatory (CTIO), La Serena, Chile; South Africa Astronomical Observatory, Sutherland, South Africa; and Siding Spring Observatory, Coonabarabran, Australia (Henderson et al. 2014). Each telescope is equipped with 9K × 9K CCDs with 10 microns pixel size. The telescopes monitor the Galactic bulge with Cousins-I filter with 10 min cadence for 8 h each night, in order to detect exoplanets near and beyond snow line via microlensing. KMTNet discovered its first microlensing exoplanet in 2015, KMT-2015-1b, a giant planet orbiting a low-mass M-dwarf (Hwang et al. 2015).

PLANET

Probing Lensing Anomalies NETwork (PLANET) was established in 1995 in order to follow-up microlensing events alerted by OGLE and MOA. The network consisted of five 1 m-class telescopes: 1.54 m Danish telescope at La Silla, Chile; 1.0 m telescope at Canopus Observatory, Tasmania, Australia; 0.6 m telescope at Perth Observatory, Perth, Australia; 1.0 m telescope at South African Astronomical Observatory, Sutherland, South Africa; 1.52 m Rockefeller telescope at Boyden Observatory, Bloemfontein, South Africa, to establish a round-the-world network for the southern hemisphere. Since 2005, PLANET collaborated with the RoboNet-1.0 team. In January 2009, PLANET merged with MicroFUN.

MicroFUN

Microlensing Follow-Up Network (MicroFUN) is an international collaboration to follow-up microlensing events in the Galactic bulge from OGLE and MOA (Gould 2008). The collaboration began in 2003. To date, more than 25 telescopes on 5 continents are used. Over half its members are amateur astronomers. In 2008, Micro-FUN along with other collaborations announced the discovery of the first multiple planetary system detected using microlensing technique, OGLE-2006-BLG-109Lb,c (Gaudi et al. 2008).

RoboNet

RoboNet is a network of robotic telescopes which follow-up microlensing events in the Galactic bulge (Tsapras et al. 2009). The network uses three 2 m telescopes: the Faulkes Telescope North at Maui, Hawaii; the Faulkes Telescope South at Siding Springs, Australia; and the Liverpool telescope at La Palma, the Canary Islands.

MiNDSTEp

Microlensing Network for the Detection of Small Terrestrial Exoplanets (MiND-STEp) is a microlensing follow-up network. MiNDSTEp consists of the 1.5 m Danish telescope at La Silla, Chile, two 1.2 m MONET telescope at McDonald Observatory and South African Astronomical Observatory and the 0.6 m Salerno University Telescope at Salerno University Observatory, Italy.

AST3

AST3 (Antarctic Schmidt Telescopes) is a project that includes in its scope the search for exoplanets with the gravitational microlensing technique and detecting transient objects, supernova and other objects with gamma-ray bursts, proposed by CCAA (Chinese Center of Antarctic Astronomy) and Texas A & M University (Cui et al. 2008; Yuan et al. 2014). For the purposes of this project, three telescopes of 50 cm aperture each were built at the Antarctic Kunlun Station near Dome A in Antarctica. Continuous, long-term monitoring will also be possible with this project.

2.2.4   Space-Based Microlensing Missions

Euclid

Euclid is an ESA observatory probe to understand dark matter and dark energy which will be launched in 2020 (Laureijs et al. 2011). *Euclid* is planned to have a diameter of 1.2 m and will have two instruments, a visual instrument (VIS: Visible and Near Infrared Imaging Channels instrument) and a near infrared instrument (NISP: Near Infrared Spectrometer and Photometer). The VIS will be equipped with 36 CCDs in broad band ($R + I + Z$) which will be used for measuring the shape of galaxies. The NISP is equipped with 16 HgCdTe NIR detectors with Y, J, H bands and a field of view of 0.55 deg^2, and a resolution better than 0.3 arcseconds. There are two primary cosmology surveys in this mission, a wide survey which covers 15, 000 deg^2 of extragalactic sky and a deep survey which covers 40 deg^2 at ecliptic poles. Additional science surveys, including a microlensing exoplanet survey, can be added (Penny et al. 2013). In this survey, *Euclid* will detect not only microlensing exoplanets, but also transiting exoplanets (Fig. 2.3, McDonald et al. 2014).

WFIRST

The WFIRST (Wide-Field Infrared Survey Telescope) is a NASA mission designed to detect exoplanets using the gravitational microlensing technique and dark energy

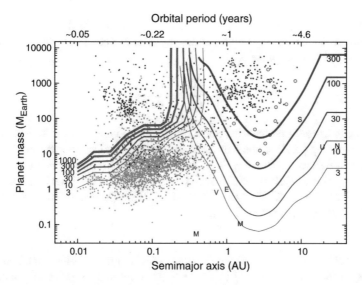

Fig. 2.3 Planet detection capability of *Euclid*. The contours indicate the number of detections using transit (red) and microlensing (blue) techniques, assuming one planet per star at each point in the planet mass-semi-major axis plane. The blue dots are published microlensing exoplanets. The red dots are published planet detections and the pink dots are the published *Kepler* candidates (McDonald et al. 2014)

with weak lensing (Spergel et al. 2013, 2015). WFIRST will be a 2.4-m obscured aperture space telescope equipped with the wide-field instrument and the coronagraph instrument. The wide-field instrument will consist of 6×3 4K $\times$ 4K HgCdTe detector array which can do both photometry (0.76–2.0 μm) and spectroscopy (1.35–1.89 μm). Microlensing observations will be performed twice yearly for 72 continuous days, with 5 days interruption each month when the Moon is near the Galactic bulge. WFIRST is expected to detect 2600 exoplanets around stars, 50 of which will be smaller than the Earth, 370 Earth-mass planets and 30 free floating planets, if there is one planet per star in the Galaxy. It can measure the mass function of cold exoplanets to better than 10% per decade in mass for planet masses >0.3 Earth-mass.

Spitzer

Spitzer is a 0.85-m NASA infrared space-based telescope launched in 2003. It is in Earth-trailing orbit with distance about 1 AU from the Earth which can be used to measure the microlensing parallax. The simultaneous observations of the same microlensing event from two distant location, such as from the Earth and *Spitzer*, provide the difference in time of peak, Δt_0, and impact parameter, Δu_0, of the event (Alcock et al. 1995; Gould et al. 2009). The microlensing parallax vector can be written as

$$\pi_E = \frac{\mathrm{AU}}{D} \left(\frac{\Delta t_0}{t_E}, \Delta u_0 \right) \qquad (2.2)$$

where t_E is Einstein crossing time (See Sect. 5.1.2) and D is the distance between two observers projected onto the plane of the sky. The radius of the projected Einstein radius on the observer plane is

$$r_E = \frac{\text{AU}}{\pi_E} = \frac{4GM_l}{\theta_E c^2} \, , \tag{2.3}$$

where θ_E is Einstein radius (See Sect. 5.1.2) and M_l is lens mass. Therefore, from the microlensing parallax, the lens mass and distance can be constrained. Due to its long distance from the Earth, *Spitzer* has been used to observe many microlensing parallaxes, such as OGLE-2005-SMC-001 (Dong et al. 2007), OGLE-2014-BLG-0124L (Udalski et al. 2015), OGLE-2014-BLG-0939 (Yee et al. 2015) and OGLE-2014-BLG-1050L (Zhu et al. 2015).

Gaia

Gaia is a ESA space satellite operated at the second Lagrange point of the Sun-Earth-Moon system. *Gaia* consists of two 1.45×0.5 m telescopes and was launched on 19 December 2013. The *Gaia* scientific goal is to perform astrometry of all objects down to magnitude 20. Scanning frequency depends on the target celestial position. There are objects that will be observed almost 200 times during its five-year main mission, while some objects are observed for only a few ten times. On average, *Gaia* monitors each target about 70 times over five-year period (Gaia Collaboration et al. 2016). Although, *Gaia* is not desired to perform microlensing survey. *Gaia* can measure microlensing and astrometric microlensing signal provided by nearby lenses (Belokurov and Evans 2002). To date, *Gaia* discovered few microlensing events, such as Gaia16aue and Gaia16aye.

References

Alcock C, Allsman RA, Alves D et al (1995) ApJ 454:L125
Alonso R, Brown TM, Torres G et al (2004) ApJ 613:L153
Alsubai KA, Parley NR, Bramich DM et al (2013) Acta Astron 63:465
Anderson DR, Hellier C, Gillon M et al (2010) ApJ 709:159
Bakos GÁ, Hartman JD, Torres G et al (2011) Eur Phys J Web Conf, vol 11, p 1002
Bakos G, Noyes RW, Kovács G, Stanek KZ, Sasselov DD, Domsa I (2004) PASP 116:266
Bakos GÁ, Noyes RW, Kovács G et al (2007) ApJ 656:552
Bakos GÁ, Csubry Z, Penev K et al (2013) PASP 125:154
Bakos G, Afonso C, Henning T et al (2009) IAU Symposium, vol 253, p 354–357
Barge P, Baglin A, Auvergne M et al (2008) A&A 482:L17
Batalha NM, Rowe JF, Bryson ST et al (2013) ApJS 204:24
Bayliss D, Gillen E, Eigmuller P et al (2017) ArXive-prints
Bean JL, Seifahrt A, Hartman H et al (2010c) ApJ 711:L19
Belokurov VA, Evans NW (2002) MNRAS 331:649
Benedict GF, McArthur BE, Bean JL et al (2010) AJ 139:1844
Benedict GF, McArthur BE, Forveille T et al (2002) ApJ 581:L115
Bennett DP, Batista V, Bond IA et al (2014) ApJ 785:155

Bond IA, Rattenbury NJ, Skuljan J et al (2002) MNRAS 333:71
Bond IA, Udalski A, Jaroszyński M et al (2004) ApJ 606:L155
Borucki WJ, Koch DG, Basri G et al (2011a) ApJ 728:117
Borucki W, Koch D, Basri G et al (2008) In: Sun Y-S, Ferraz-Mello S, Zhou J-L (eds) Exoplanets: detection, formation and dynamics, IAU Symposium, vol 249, pp 17–24
Broeg C, Fortier A, Ehrenreich D et al (2013) Eur Phys J Web Conf 47. Article ID 03005
Brown T (2013) Publications de l'Observatoire Astronomique de Beograd 92:91
Butters OW, West RG, Anderson DR et al (2010) A&A 520:L10
Cassan A, Kubas D, Beaulieu J-P et al (2012) Nature 481:167
Charbonneau D, Brown TM, Latham DW, Mayor M (2000) ApJ 529:L45
Chauvin G, Lagrange A-M, Dumas C et al (2004) A&A 425:L29
Chauvin G, Lagrange A-M, Dumas C et al (2005) A&A 438:L25
Collier Cameron A, Guenther E, Smalley B et al (2010) MNRAS 407:507
Coughlin JL, Mullally F, Thompson SE et al (2016) ApJS 224:12
Crouzet N, McCullough PR, Long D et al (2017) AJ 153:94
Cui X, Yuan X, Gong X (2008) Society of photo-optical instrumentation engineers (SPIE) conference series, vol 7012
Díaz RF, Almenara JM, Bonomo A et al (2011) EPSC-DPS Joint Meeting, p 1243
Dominik M, Albrow MD, Beaulieu J-P et al (2002) Planet Space Sci 50:299
Dong S, Udalski A, Gould A et al (2007) ApJ 664:862
Gaia Collaboration, Prusti T, de Bruijne JHJ et al (2016) A&A 595:A1
Gaudi BS (2012) ARA&A 50:411
Gaudi BS, Bennett DP, Udalski A et al (2008) Science 319:927
Gillon M, Jehin E, Lederer SM et al (2016) Nature 533:221
Gillon M, Triaud AHMJ, Demory B-O et al (2017) Nature 542:456
Gillon M, Jehin E, Fumel A, Magain P, Queloz D (2013) Eur Phys J Web Conf 47. Article ID 03001
Gillon M, Jehin E, Magain P et al (2011) Eur Phys J Web Conf 11. Article ID 06002
Gould A (2008) Manchester Microlensing Conference, 38
Gould A, Udalski A, Monard B et al (2009) ApJ 698:L147
Gould A, Loeb A (1992) ApJ 396:104
Hartman JD, Bakos GÁ, Torres G et al (2011) ApJ 742:59
Heintz WD (1978) ApJ 220:931
Henderson CB, Gaudi BS, Han C et al (2014) ApJ 794:52
Henderson CB, Poleski R, Penny M et al (2016) PASP 128(12):124401
Henry GW, Marcy GW, Butler RP, Vogt SS (2000) ApJ 529:L41
Howell SB, Sobeck C, Haas M et al (2014) PASP 126:398
Hwang K-H, Han C, Choi J-Y et al (2015) arXiv: 1507.05361
Jenkins JM, Twicken JD, Batalha NM et al (2015) AJ 150:56
Kennedy GM, Kenyon SJ (2008) ApJ 682:1264
Koch DG, Borucki WJ, Basri G et al (2010) ApJ 713:L79
Koppenhoefer J, Saglia RP, Fossati L et al (2013) MNRAS 435:3133
Kuhn RB, Rodriguez JE, Collins KA et al (2016) MNRAS 459:4281
Laureijs R, Amiaux J, Arduini S, et al (2011) Euclid definition study report, ESA
Lecar M, Podolak M, Sasselov D, Chiang E (2006) ApJ 640:1115
Léger A, Rouan D, Schneider J et al (2009) A&A 506:287
Lin DNC (2008) Sci Am 298:50
Mandushev G, Quinn SN, Buchhave LA et al (2011) ApJ 741:114
Mao S, Paczynski B (1991) ApJ 374:L37
Marois C, Macintosh B, Barman T et al (2008) Science 322:1348
Martioli E, McArthur BE, Benedict GF, Bean JL, Harrison TE, Armstrong A (2010) ApJ 708:625
Masciadri E, Raga A (2004) ApJ 611:L137
Mayor M, Queloz D (1995) Nature 378:355
McCullough PR, Stys JE, Valenti JA, Fleming SW, Janes KA, Heasley JN (2005) PASP 117:783

McCullough PR, Stys JE, Valenti JA et al (2006) ApJ 648:1228
McDonald I, Kerins E, Penny M et al (2014) MNRAS 445:4137
Morton TD, Bryson ST, Coughlin JL et al (2016) ApJ 822:86
Moutou C, Deleuil M (2015) arXiv:1510.01372
Mróz P, Udalski A, Skowron J et al (2017) Nature 548:183
Nielsen EL, Close LM, Biller BA, Masciadri E, Lenzen R (2008) ApJ 674:466
Penny MT, Kerins E, Rattenbury N et al (2013) MNRAS 434:2
Pepe F, Molaro P, Cristiani S et al (2014) Astron Nachr 335:8
Pepper J, Gould A, Depoy DL (2003) Acta Astron 53:213
Pepper J, Pogge RW, DePoy DL et al (2007) PASP 119:923
Pepper J, Kuhn RB, Siverd R, James D, Stassun K (2012) PASP 124:230
Pollacco DL, Skillen I, Collier Cameron A et al (2006) PASP 118:1407
Pravdo SH, Shaklan SB (2009) ApJ 700:623
Rauer H, Catala C, Aerts C et al (2014) Exp Astron 38:249
Reuyl D, Holmberg E (1943) ApJ 97:41
Ricker GR, Winn JN, Vanderspek R et al (2014) Space telescopes and instrumentation: optical.
    infrared, and millimeter wave 9143:914320
Sahu KC, Casertano S, Bond HE et al (2006) Nature 443:534
Sasselov DD (2003) ApJ 596:1327
Shporer A, Brown T, Lister T et al (2011) In: Sozzetti A, Lattanzi MG, Boss AP (eds) The astro-
    physics of planetary systems: formation, structure, and dynamical evolution, IAU Symposium,
    vol 276, pp 553–555
Siverd RJ, Beatty TG, Pepper J et al (2012) ApJ 761:123
Smith AMS et al (2014) Contrib Astron Obs Skaln Pleso, vol 43, p 500
Spergel D, Gehrels N, Baltay C et al (2015) arXiv:1503.03757
Spergel D, Gehrels N, Breckinridge J et al (2013) arXiv:1305.5422
Strand KA (1943) PASP 55:29
Sumi T (2010) Pathways towards habitable planets. In: Coudé Du Foresto V, Gelino DM, Ribas I
    (eds) Astronomical society of the pacific conference series, vol. 430, p 225
Sumi T, Kamiya K, Bennett DP et al (2011) Nature 473:349
Sumi T, Bennett DP, Bond IA et al (2013) ApJ 778:150
Swift JJ, Bottom M, Johnson JA et al (2015) J Astron Telesc Instrum Syst 1(2):027002
Talens GJJ, Spronck JFP, Lesage A-L et al (2017b) A&A 601:A11
Talens GJJ, Albrecht S, Spronck JFP et al (2017a) A&A 606:A73
Triaud AHMJ, Collier Cameron A, Queloz D et al (2010) A&A 524:A25
Tsapras Y, Street R, Horne K et al (2009) Astron Nachr 330:4
Udalski A (2009) The Variable Universe: A Celebration of Bohdan Paczynski. In: Stanek KZ (ed)
    Astronomical Society of the Pacific Conference Series, vol 403, p 110
Udalski A, Szymanski M, Mao S et al (1994) ApJ 436:L103
Udalski A, Kubiak M, Szymanski M (1997) Acta Astron 47:319
Udalski A, Paczynski B, Zebrun K et al (2002) Acta Astron 52:1
Udalski A, Szymanski MK, Soszynski I, Poleski R (2008) Acta Astron 58:69
Udalski A, Yee JC, Gould A et al (2015) ApJ 799:237
Wheatley PJ, Pollacco DL, Queloz D et al (2013) Eur Phys J Web Conf 47. Article ID 13002
Wolszczan A, Frail DA (1992) Nature 355:145
Wyrzykowski Ł, Rynkiewicz AE, Skowron J et al (2015) ApJS 216:12
Yee JC, Udalski A, Calchi Novati S et al (2015) ApJ 802:76
Yuan X, Cui X, Gu B et al (2014) Ground-based and Airborne telescopes V, vol 9145, 91450F
Zhu W, Udalski A, Gould A et al (2015) ApJ 805:8

Chapter 3
Transit Timing Variation and Transmission Spectroscopy Analyses of the Hot Neptune GJ3470b

The transit method is one of the most effective exoplanet detection method, detecting more than 2,700 exoplanets, including over 2,300 by *Kepler* (Morton et al. 2016). The transit method can detect planets ranging in size from Earth to larger than Jupiter. The transit timing variation (TTV) method has been used to find at least 10 additional exoplanets and a hundred of candidates (Agol et al. 2005; Holman and Murray 2005; Holman et al. 2010; Ford et al. 2012a, b; Fabrycky et al. 2012; Steffen et al. 2012a, b, 2013; Mazeh et al. 2013).

In addition to the discovery of new exoplanets, characterization of planetary interiors and atmospheres is a rapidly developing area. One method that is used to study planetary atmospheres is transmission spectroscopy, which measures the variation in transit depth with wavelength (Seager and Deming 2010). This method has been applied to several transiting exoplanets [e.g. HD189733b (Grillmair et al. 2008; Swain et al. 2010), GJ1214b (Bean et al. 2010a; Kreidberg et al. 2014) and GJ436b (Knutson et al. 2014)].

GJ3470b is a hot Neptune exoplanet orbiting an M dwarf and the first sub-Jovian planet to exhibit Rayleigh scattering. In this Chapter, we present transit timing variation and transmission spectroscopy analyses of multi-wavelength optical photometry from telescopes at the Thai National Observatory, and the 0.6-m PROMPT-8 telescope in Chile (Awiphan et al. 2016a).

3.1 Characterizing Transiting Exoplanet

3.1.1 Probability of Transits

A planetary transit occurs when a planet moves in front of its host star and blocks a fraction of stellar light, causing the dimming of the stellar brightness. In order to

© Springer International Publishing AG, part of Springer Nature 2018
S. Awiphan, *Exomoons to Galactic Structure*, Springer Theses,
https://doi.org/10.1007/978-3-319-90957-8_3

observe transiting exoplanets, only systems with nearly edge-on orbit can be observed using this technique. Therefore, only a small fraction of all planets can be observed with the transit technique from the Earth.

The probability of detecting transit events (P_{tran}) relates to the host star's radius and the distance between star and planet (Borucki and Summers 1984). Since planetary systems in the Galaxy are likely to be randomly oriented, the probability of detecting transiting exoplanets is equal to the ratio of the solid angle in which planets will be seen to transit to the total solid angle, 4π, For a system with elliptical orbit, the geometric probability of transit can be written as,

$$P_{tran} = \left(\frac{R_* + R_p}{a_p} \right) \left(\frac{1 + e_p \cos(\pi/2 - \omega_p)}{1 - e_p^2} \right) , \qquad (3.1)$$

where e_p is the orbital eccentricity of the planet and ω_p is the planet argument of periastron (Charbonneau et al. 2007). However, the orbital parameters e_p and ω are not known for all planets. As more than 2,700 transiting exoplanet have been discovered to date, most of them are short orbital period exoplanets, due to their higher detection probabilities.

3.1.2　Transiting Exoplanet Light Curves

The transiting exoplanet light curve is described by the total measured flux of star and planet,

$$F = F_p + F_* - \begin{cases} 0 & \text{outside eclipses ,} \\ F_* \alpha_{\text{tra}} & \text{transits ,} \\ F_p \alpha_{\text{occ}} & \text{occultations ,} \end{cases} \qquad (3.2)$$

where F_p and F_* are the flux of planet and star, respectively. α is the fraction of overlap area between planet and star during transit. The shape of transit dip in transit light curve is described by three main parameters: transit depth, transit duration, and ingress/egress duration. These parameters depend on the sizes of star and planet, orbital inclination and separation of the planet (Fig. 3.1).

Transit Depth

The relative depth of a transit is given by the ratio of projection area of the planet radius, R_p, and the star radius, R_*,

$$\delta = \left(\frac{R_p}{R_*} \right)^2 \left(1 + \frac{F_p}{F_*} \right) . \qquad (3.3)$$

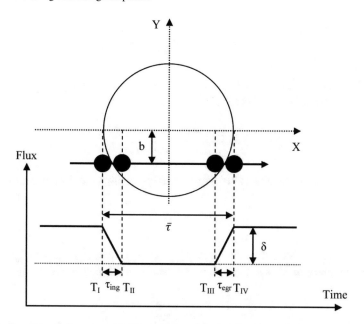

Fig. 3.1 Transiting light curve with four contact points and three main parameters: transit depth, transit duration, and ingress and egress durations. Adapted from Winn (2010)

The planetary flux is very faint compared to the stellar flux. Therefore, the planet flux is usually assumed to be negligible:

$$\delta \approx \left(\frac{R_p}{R_*} \right)^2 . \tag{3.4}$$

Thus, the measurement of transit depth can provide the radius of the exoplanet. The stellar and planetary fluxes are assumed to be constant, however, in reality, the stellar variability might cause a significant change in the transit light curve (Czesla et al. 2009; Silva 2003). Transit light curves are also not flat-bottomed, due to limb darkening effect, which causes a fainter surface brightness near the limb of the star. In general, limb darkening is described using a quadratic law where the intensity, I, of the star at a point (X,Y) in Fig. 3.1 is

$$I \propto 1 - c_1(1 - \sqrt{1 - X^2 - Y^2}) - c_2(1 - \sqrt{1 - X^2 - Y^2})^2 , \tag{3.5}$$

where c_1 and c_2 are quadratic limb darkening coefficients (See Sect. 3.1.4).

Transit Duration

The transit duration is the time between the first and last contact points of the transit. For a circular orbit, the duration is defined by,

$$\bar{\tau} = T_{IV} - T_I = \frac{P_p}{\pi} \arcsin\left(\sqrt{\frac{1 - b^2}{(a_p/R_*)^2 - b^2}}\right), \quad (3.6)$$

where b is impact parameter, the minimum sky-projected distance between star and planet in the units of stellar radius,

$$b = \frac{a_p \cos i}{R_*} \frac{1 - e^2}{1 + e \sin \omega_p}. \quad (3.7)$$

For a circular orbit, the value depends only on the orbital inclination and the planet-star distance:

$$b = \frac{a_p}{R_*} \cos i. \quad (3.8)$$

From the Kepler's third law, $P^2 \propto a^3/(M_* + M_p)$, if the system has edge-on orbit and the planetary mass is neglected, the transit duration may be written as,

$$\bar{\tau} = \frac{R_*}{a_p} \frac{P_p}{\pi} = \left(\frac{3}{G\pi^2} \frac{P}{\rho_*}\right)^{\frac{1}{3}}, \quad (3.9)$$

where ρ_* is the mean density of the star. Therefore, if we can measure the transit duration, the mean stellar density can be estimated (Seager and Mallén-Ornelas 2003; Perryman et al. 2005). In the case of a planet with an eccentric orbit, the transit duration formula is much more complicated. Many theoretical models have been proposed, such as Tingley and Sackett (2005) and Winn (2010).

Ingress and Egress Durations

The shape of the transit light curve is also defined by ingress (τ_{ing}), and egress (τ_{egr}) durations which are the timing between first and second contacts, and third and fourth contacts, respectively. The ingress and egress durations are equal for a planet with a circular orbit:

$$\tau = \tau_{ing} = \tau_{egr} = T_{II} - T_I = T_{IV} - T_{III} = \bar{\tau} \frac{R_p}{R_*} \sqrt{1 - b^2}. \quad (3.10)$$

However, τ_{ing} and τ_{egr} are unequal for an eccentric orbit, because of the variation in projected speed (Winn 2010). For an edge-on circular system, Southworth et al. (2007) suggests that the transit ingress and egress durations can be use to estimate the planetary surface gravity, g_p, by using Kepler's third law and radial velocity observation (Eq. 2.1),

$$\tau = \tau_{ing} = \tau_{egr} = \frac{R_p}{a_p} \frac{P_p}{\pi} = \sqrt{\frac{2}{\pi} \frac{P_p K_*}{g_p}}. \quad (3.11)$$

3.1.3 Determining System Parameters

One of the main goals of observing transiting exoplanets is to determine the host star's and planet's physical parameters. However, the transit technique can provide only some orbital parameters of the planet. In order to obtain all parameters, radial-velocity data and some stellar parameters are required. In this section, the calculations of some planetary parameters from transit data are provided.

Impact Parameter

The impact parameter, b, is sky-projected distance between the centre of the star's disk and the centre of the planet's disk at conjunction. The impact parameter can be written in term of $\bar{\tau}$ and τ in the case of circular orbit system,

$$b^2 = \frac{(1 - \frac{R_p}{R_*})^2 - \left(\frac{\bar{\tau}-2\tau}{\bar{\tau}}\right)^2 (1 + \frac{R_p}{R_*})^2}{1 - \left(\frac{\bar{\tau}-2\tau}{\bar{\tau}}\right)^2} . \tag{3.12}$$

For a system with small planet and $\tau \ll \bar{\tau}$, the impact parameter can be approximated as,

$$b^2 = 1 - \frac{R_p}{R_*} \frac{\bar{\tau}}{\tau} . \tag{3.13}$$

Scaled Stellar Radius

The scaled stellar radius, R_*/a, is the ratio between stellar radius and semi-major axis of planetary orbit. For non-grazing transits and the limit $R_p \ll R_* \ll a$, the scaled stellar radius may be written as,

$$\frac{R_*}{a} = \frac{\pi \sqrt{\bar{\tau}\tau - \tau^2}}{P} \sqrt{\frac{R_*}{R_p}} \left(\frac{\sqrt{1 + e \sin \omega}}{\sqrt{1 - e^2}}\right) , \tag{3.14}$$

and, in the small-planet limit,

$$\frac{R_*}{a} = \frac{\pi \sqrt{\bar{\tau}\tau}}{P} \sqrt{\frac{R_*}{R_p}} \left(\frac{\sqrt{1 + e \sin \omega}}{\sqrt{1 - e^2}}\right) . \tag{3.15}$$

Mean Stellar Density

The scaled stellar radius can be used to determine a combination of the stellar mean density, ρ_*, and planetary mean density, ρ_p:

$$\rho_* + \left(\frac{R_P}{R_*}\right)^3 \rho_p = \frac{3\pi}{GP^2} \left(\frac{a}{R_*}\right)^3 . \tag{3.16}$$

In general cases, the planet-star radius ratio is small, so the stellar mean density can be obtained from the transit light curve:

$$\rho_* = \frac{3\pi}{GP^2} \left(\frac{a}{R_*}\right)^3 . \tag{3.17}$$

3.1.4 Limb-Darkening

The effect of the limb-darkening, in which the star is brighter in the middle and fainter at the edge, affects the shape of transit light curve by rounding the boxy transit light curve profile, because the declined flux during the transit is larger than the flux of a uniform source near the centre of transit and smaller near the edge. The change in brightness due to the limb darkening effect depends on the opacity of the stellar atmosphere and is strongly wavelength dependent. In order to quantify the effect, many limb-darkening laws have been advocated:

- Linear law

$$\frac{I(\mu)}{I_0} = 1 - u(1 - \mu) , \tag{3.18}$$

- Quadratic law

$$\frac{I(\mu)}{I_0} = 1 - a(1 - \mu) - b(1 - \mu)^2 , \tag{3.19}$$

- Square root law

$$\frac{I(\mu)}{I_0} = 1 - c(1 - \mu) - d(1 - \sqrt{\mu}) , \tag{3.20}$$

- Logarithmic law

$$\frac{I(\mu)}{I_0} = 1 - e(1 - \mu) - f\mu \ln(\mu) , \tag{3.21}$$

where I_0 is the specific intensity at the centre of the disk. u, a, b, c, d, e and f are the limb-darkening coefficients. The quantity μ is defined by $\mu = \cos(\gamma) = \sqrt{(1 - r^2)} = \sqrt{1 - X^2 - Y^2}$ (Fig. 3.1), where γ is the angle between the line of sight and the emergent intensity and $0 \leqslant r \leqslant 1$ is the normalized radial coordinate on the disk of the star. Claret (2000) proposed a nonlinear limb-darkening law that fits a wide range of stellar models:

$$\frac{I(\mu)}{I_0} = 1 - \sum_{n=1}^{4} c_n (1 - \mu^{n/2}) , \tag{3.22}$$

where c_n is the nonlinear limb-darkening coefficient. This effect can cause inaccuracies in planetary parameter measurements from the transit light curve. On the other hand, the limb-darkening effect in the transit light curves can provide information about host stars opacities (Knutson et al. 2007).

3.1.5 Transit Timing Variations

Transit timing variation (TTV) is a change in the transit ephemeris due to a change in the planetary orbital phase caused by the gravitational interactions of third bodies in the system, such as exoplanets or exomoons. In an idealized single transiting exoplanet system, the transit occurs at a constant interval which equals the planetary orbital period. But, in a multiple planet system, gravitational interactions between planets can cause a periodic change in orbital period, resulting in a non-zero TTV. The TTV can be used to detect additional planet, although the additional planets' transits may not be detected. However, the presence of TTV signals can be caused by many other potential scenarios, such as orbital precession and orbital decay.

For a TTV caused by a second planet, the amplitude of the TTV signal depends on the perturber's mass, which can be even lower Earth-mass, and the orbital period of the transiting planet (Holman and Murray 2005; Agol et al. 2005). The amplitude can be increased, if there is a mean-motion resonance (MMR) between planet orbital periods (Agol et al. 2005). An Earth-mass exoplanet in 2:1 resonance with a 3 day orbital period hot Jupiter would cause a 3 minute TTV signal. In 2010, Holman et al. (2010) announced the first successful exoplanet TTV signal detection in the Kepler-9 system. To date, at least 10 exoplanets have been discovered using this method.

The TTV detection requires good phase coverage and a long uninterrupted baseline of measurements which is rarely done from ground-based observations due to weather and daylight. Veras et al. (2011) suggested the number of consecutive transit observations necessary in order to characterize the perturbing planet is ≥ 50 observations, which could be done with *Kepler* (Mazeh et al. 2013). Even if the TTV curve is well sampled, it is still difficult to measure the mass of perturbing planet (Nesvorný and Morbidelli 2008; Boué et al. 2012). However, Jontof-Hutter et al. (2015) measured the mass of Kepler-138b, a Mars-mass exoplanet, using TTV by assuming co-planarity and mutual inclination of the system.

To date, no large TTV signals have been detected from ground-based surveys, but *Kepler* detected a few high amplitude TTVs (e.g Holman et al. 2010; Ballard et al. 2011; Carter et al. 2012; Lissauer et al. 2013; Jontof-Hutter et al. 2014). The discoveries of non-transiting exoplanets from *Kepler* provide useful information on the nature of the planets. Steffen et al. (2012b) used six quarters of *Kepler* data to search for planets orbiting near hot Jupiters ($1 \geq P_p \geq 5$ days), but none of 63 hot Jupiter candidates shows evidence of a TTV signal. However, 5 of 31 warm Jupiters ($6.3 \geq P_p \geq 15.8$ days) show evidence of TTVs in that study.

3.1.6 Transmission Spectroscopy

During a transit, the light from the host star is filtered through the upper atmosphere of the planet and a portion of the light is absorbed. The absorption is wavelength

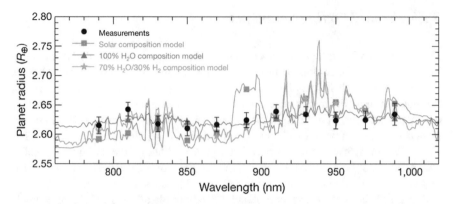

Fig. 3.2 The measurements (black circles) and theoretical transmission spectrum of GJ1214b for atmospheres with a solar composition (orange squares), a 100% water vapour composition (blue triangles) and a mixed composition of 70% water vapour and 30% molecular hydrogen by mass (green stars) (Bean et al. 2010a)

dependent due to the scattering properties of elements and molecules in the planetary atmosphere (Fig. 3.2). At the wavelengths with high atmospheric absorption, the planet appears larger than the observations at wavelengths with low atmospheric absorption, due to the additional blocking area from opaque atmosphere:

$$\frac{R_p(\lambda)}{R_p(\lambda_0)} = \sqrt{\frac{\Delta F}{F}(\lambda)}, \tag{3.23}$$

where $\frac{\Delta F}{F}(\lambda)$ is relative flux drop (Seager and Sasselov 2000; Brown 2001; Hubbard et al. 2001; Hui and Seager 2002). Therefore, by observing transits and determining the exoplanet's radius at multiple wavelengths, the absorption spectrum and the composition of the planetary atmosphere can be inferred (Swain et al. 2008). This method is called transmission spectroscopy.

For an ideal gas atmosphere in hydrostatic equilibrium with a cloudless atmosphere, the planetary pressure changes with the altitude as $d \ln(p) = -\frac{1}{h} dz$, where h is the atmospheric scale height:

$$h = \frac{k_B T_p}{\mu g_p}, \tag{3.24}$$

with k_B the Boltzmann constant, T_p the planet's temperature, μ the mean molecular mass of planet atmosphere and g_p the planetary surface gravity. Without any atomic or molecular absorption, a scaling law for the scattering cross section, σ, is assumed to be $\sigma = \sigma_0(\lambda/\lambda_0)^\alpha$, where α is a scaling factor ($\alpha = -4$ for Rayleigh scattering), as in Lecavelier Des Etangs et al. (2008). Therefore, the slope of the planetary radius as a function of wavelength can be express as $dR_p/d \ln \lambda = \alpha h$. From the atmospheric scale height relation, the mean molecular weight of the planetary atmosphere can be estimated as

$$\mu = \alpha T_p k_b \left(g_p \frac{\mathrm{d} R_p}{\mathrm{d} \ln \lambda} \right)^{-1} . \tag{3.25}$$

For gas giant planets, their atmospheres are defined by the region above the altitude where the planet's atmosphere becomes optically thick and absorbs all light at all wavelengths. Therefore, the presence of cloud in the planet's atmosphere can prevent radiation from being transmitted deep into the planet, which can cause a shallower atmospheric scale height than a cloud-free atmosphere. In the case of a cloudless atmosphere, absorption lines from volatile molecules at near-infrared wavelengths should be detectable.

On the other hand, an exoplanet with low atmospheric H/He abundance provides a smaller scale height and flatter transmission spectrum. However, the presence of high-altitude hazes might cause difficulty in separating between H/He-rich and volatile-rich envelopes, because the haze can hide molecules in the lower atmosphere and produce near-infrared transmission spectra dominated by Mie scattering (Howe and Burrows 2012).

To date, *Kepler* has discovered more than 4,000 planetary candidates (Coughlin et al. 2016). Most of them are super-Earth and Neptune-sized planet candidates (1.25 $- 6\,R_{\oplus}$) which confirms the large fraction of small planets, super-Earth and Neptune-like exoplanets within the exoplanet population (Howard et al. 2012; Coughlin et al. 2016). Exoplanets within this radius range likely comprise solid core (super-Earths), H/He gas and volatile envelope (Neptune-like exoplanets).

In order to classify the transition between super-Earths and Neptune-like exoplanets, Lopez and Fortney (2014) suggested that planets with radius larger than about 1.75 $R_{\oplus}$ have H/He envelopes. Nevertheless, the transition point may vary between 1.5 and 2.0 $R_{\oplus}$ (Weiss and Marcy 2014; Marcy et al. 2014). Instead of classification by planetary radii, Rafikov (2011) studied envelope accretion of Neptune-like planets and suggested a mass transition limit at 10 $M_{\oplus}$ or larger for close-in planets. Piso et al. (2015) also suggested that the minimum core mass to form a Neptune-like planet is $\sim$8 $M_{\oplus}$ at 5 AU and $\sim$5 $M_{\oplus}$ at 100 AU. However, in the case of low-density super-Earth-sized exoplanets, the mass of the planet alone cannot be used to be a classification [e.g. Kepler-11f (Fig. 3.3, Lissauer et al. 2011) and Kepler-51b (Masuda 2014)]. Planetary average densities are also unable to confirm the predicted transition between super-Earths and Neptune-like exoplanets, due to the broad range of detected planet densities which overlap the transition range (Howe et al. 2014).

One feature that can be used to determine the core-to-envelope transition regime is the amount of hydrogen and helium in the planet envelope (Miller-Ricci and Fortney 2010). An exoplanet with H/He-rich atmosphere has a large atmospheric scale height due to its small mean molecular weight. Therefore, transmission spectroscopy studies of a super-Earth and Neptune-sized planet provide not only information on the planetary atmospheres, but also its classification between super-Earths and Neptune-like exoplanets.

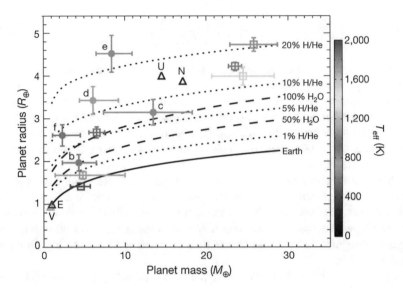

Fig. 3.3 Mass-radius relationship of exoplanets in Kepler-11 system (filled circles with labels), other transiting exoplanet (open squares): in order of ascending radius, Kepler-10b, CoRoT-7b, GJ1214b, Kepler-4b, GJ436b and HAT-P-11b, and planets in the Solar system (open triangles): Venus (V), Earth (E), Neptune (N) and Uranus (U). The colours of the points show planetary temperatures and lines represent models of planets (Lissauer et al. 2011)

3.2 Photometry Pipeline

The photometric study of transiting exoplanets requires very precise analysis due to the small variation of the transit light curve. We perform photometric analysis using Python photometry scripts. The method of photometric reductions are shown in this section.

Signal Calculation

To obtain the brightness of an object, the signal count in a certain area on the CCD is used. The count can be defined as summation of the signal inside a circular aperture. As some pixels are at the edge of the aperture, the signal, S, at that pixel is weighted by the fraction of the aperture area in that pixel.

Sky Background Calculation

The sky is not completely dark, due to the light from both nearby stellar sources and scattered light from Earth sources. The background level of each object is calculated from the mode of the signal inside an annulus around the object. The mode of the background is estimated from the formula (Da Costa 1992),

$$\text{mode} = 3 \times \text{median} - 2 \times \text{mean} . \tag{3.26}$$

In following work, the pixels with signal larger than 3-σ variation are also rejected.

Centroid Calculation

In order to obtain full transit light curves, observations must be taken over hours. As the telescope might not be fully-guided, the positions of object in the images will shift a few pixels during the observations. Therefore, the centroid point of each object in each image must be calculated in order to get the precise position of the object. In the script, we provide two methods to calculate the centroid of the star.

- **Maximum Light Method**

In this method, we calculated the sum of the signal in the defined aperture. The centre of the apertures was shifted by a defined step size within a square box region. The position that provides the maximum signal is taken to be the centroid of the object in that image.

- **Image Centroiding Method**

In the image centroiding method, the centroid is calculated using the centroid position of the signal on both axis (Da Costa 1992). For a box of size $2a \times 2a$, the summation of signal, S, in both x and y axis are

$$X_i = \sum_{j=-a}^{a} S_{i,j} \tag{3.27}$$

and

$$Y_j = \sum_{i=-a}^{a} S_{i,j} \, , \tag{3.28}$$

where i and j is the x and y position in the box. The mean intensity in x and y are

$$\bar{X} = \frac{1}{2a+1} \sum_{i=-a}^{a} X_i \tag{3.29}$$

and

$$\bar{Y} = \frac{1}{2a+1} \sum_{j=-a}^{a} Y_j \, . \tag{3.30}$$

For the positions where the signal is above the mean density, the centroid is defined by,

$$X = \frac{\sum_{i=-a}^{a(X_i \geq \bar{X})} (X_i - \bar{X}) x_i}{\sum_{i=-a}^{a(X_i \geq \bar{X})} (X_i - \bar{X})} \, , \tag{3.31}$$

and

$$Y = \frac{\sum_{j=-a}^{a(Y_j \geq \bar{Y})}(Y_j - \bar{Y})y_j}{\sum_{j=-a}^{a(Y_j \geq \bar{Y})}(Y_j - \bar{Y})} \, ,$$

(3.32)

where x and y is the x and y position in the image.

3.3 Hot-Neptune GJ3470b

An exoplanet around a nearby M dwarf is favourable for transmission spectroscopic studies, due to its large planet-host radius ratio. GJ3470b was first discovered with the HARPS spectrograph and confirmed with follow-up transit observations with the TRAPPIST, Euler and NITES telescopes (Bonfils et al. 2012). GJ3470b is a good target for transmission spectroscopy because it has a large change in transit depth with wavelength due to its large atmospheric opacity (Bento et al. 2014). GJ3470b is also the first sub-Jovian planet that shows a significant Rayleigh scattering slope (Nascimbeni et al. 2013). To date, GJ3470b has been observed at several optical and near-infrared wavelengths (Fukui et al. 2013; Crossfield et al. 2013; Demory et al. 2013; Nascimbeni et al. 2013; Biddle et al. 2014; Ehrenreich et al. 2014; Dragomir et al. 2015).

Fukui et al. (2013) observed GJ3470b with simultaneous optical and near-infrared observations with the 0.5 m MITSuME and 1.88 m telescopes at Okayama Astrophysical Observatory. They suggested that GJ3470b has a cloud-free atmosphere. Nascimbeni et al. (2013) combined their optical observations with the Large Binocular Telescope (LBT) (Demory et al. 2013; Fukui et al. 2013). Their result suggests that the GJ3470b atmosphere is cloud-free with a high-altitude haze of tholins. They also found a strong Rayleigh-scattering slope at visible wavelengths.

However, Crossfield et al. (2013) performed an observation with the MOSFIRE spectrograph at the Keck I telescope. They concluded that the GJ3470b atmosphere yields a flat transmission spectrum which indicates methane-poor, metal-rich, optically-thick clouds or a hazy atmosphere. Biddle et al. (2014) presented 12 new broad-band optical transit observations and concluded that GJ3470b has a hydrogen-rich atmosphere exhibiting a strong Rayleigh-scattering slope from a hazy atmosphere with 50 times solar abundance.

A recent study by Ehrenreich et al. (2014) with the Wide Field Camera-3 (WFC3) on the *Hubble* Space Telescope (HST) in the near-infrared also suggested that GJ3470b is dominated by a cloudy hydrogen-rich atmosphere with extremely low water volume mixing ratios (<1 ppm). Dragomir et al. (2015) provided shorter-wavelength transmission spectroscopic results with the LCOGT network and the Kuiper telescope in order to verify the Rayleigh scattering signal in the 400–900 nm region. They found a strong Rayleigh scattering slope that indicates a H/He atmosphere with hazes as in previous studies.

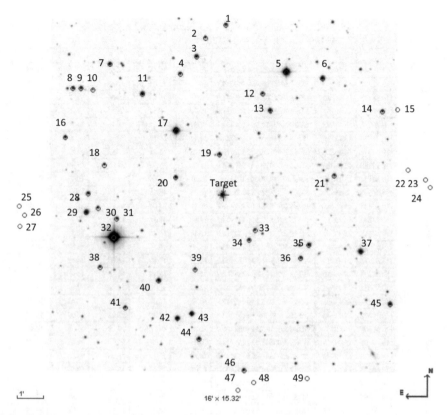

Fig. 3.4 Finder chart of GJ3470 and reference stars generated from *Aladin* (See http://aladin.u-strasbg.fr/)

GJ3470b is also a good target to observe with small telescopes, as it is a nearby M-dwarf with $I \approx 12$ and has many bright reference stars (Table 3.1). Moreover, the system is located in an uncrowded field that is good for small telescopes with large pixel size (Fig. 3.4).

3.4 Observations and Data Analysis

3.4.1 Photometric Observations

Photometric observations of exoplanet GJ3470b were conducted between December 2013 and March 2016. We obtained 10 transits, including 6 full transits and 4 partial transits. The UT date of the mid-transit time, instrument, filter, exposure time and

Table 3.1 GJ3470 and reference stars from Fig. 3.4 locations and magnitudes based on Sloan digital sky survey III, catalogue DR10 (Ahn et al. 2014)

	RA	Dec	Name	u'	g'	r'	i'	z'
GJ3470	119.77446	15.39152	SDSS J075905.77+152328.8	15.93	15.44	11.87	10.76	10.58
1	119.77278	15.49438	SDSS J075905.46+152939.7	16.03	15.74	14.26	14.93	14.32
2	119.78615	15.48679	SDSS J075908.69+152911.5	16.77	15.60	15.03	14.79	14.67
3	119.79169	15.47531	SDSS J075910.02+152831.0	15.40	13.58	13.09	13.02	13.70
4	119.80238	15.46466	SDSS J075912.54+152752.2	17.37	15.58	15.49	14.63	14.48
5	119.73375	15.46569	SDSS J075856.09+152756.5	11.44	10.34	9.85	9.74	10.74
6	119.71049	15.46216	SDSS J075850.52+152743.7	16.25	13.70	12.55	12.53	13.92
7	119.84758	15.47029	SDSS J075923.41+152812.9	16.94	15.62	13.86	14.19	13.75
8	119.87124	15.45570	SDSS J075929.10+152720.3	16.08	14.69	14.15	16.17	13.86
9	119.86605	15.45583	SDSS J075927.85+152720.8	16.33	14.39	13.67	16.66	13.29
10	119.85844	15.45485	SDSS J075926.01+152717.3	20.17	17.65	16.21	15.44	15.04
11	119.82647	15.45255	SDSS J075918.34+152709.2	14.60	13.18	12.84	12.72	13.05
12	119.74911	15.45280	SDSS J075859.78+152710.0	16.54	15.20	14.73	14.52	14.40
13	119.74429	15.44268	SDSS J075858.62+152633.6	15.93	14.69	13.62	13.54	14.06
14	119.67215	15.44238	SDSS J075841.32+152632.5	16.01	14.89	16.20	14.65	14.61
15	119.66215	15.44350	SDSS J075838.89+152636.5	17.22	15.71	15.15	14.96	14.89
16	119.87621	15.42603	SDSS J075930.30+152533.0	18.23	15.70	14.29	13.84	13.12
17	119.80484	15.43000	SDSS J075913.15+152547.7	14.84	11.34	10.41	10.07	10.23
19	119.77672	15.41596	SDSS J075906.40+152457.4	15.71	14.37	13.92	15.99	13.72
20	119.80505	15.40192	SDSS J075913.21+152406.9	16.74	14.54	13.79	13.91	13.41
21	119.70301	15.40353	SDSS J075848.72+152412.4	16.31	15.02	15.69	14.47	14.40
22	119.65524	15.40720	SDSS J075837.25+152425.9	16.39	15.15	15.66	14.62	14.55
23	119.64390	15.40111	SDSS J075834.53+152403.7	15.35	15.43	15.39	12.57	13.65
24	119.64075	15.39658	SDSS J075833.78+152347.4	16.14	14.99	15.60	15.08	14.41
25	119.90595	15.38453	SDSS J075937.42+152304.1	17.91	15.96	15.23	14.97	14.83
26	119.90222	15.37913	SDSS J075936.53+152244.7	16.02	14.67	14.31	16.86	14.14
27	119.90544	15.37202	SDSS J075937.30+152219.1	16.11	14.96	14.63	14.52	14.51
28	119.86138	15.39209	SDSS J075926.74+152331.4	15.69	14.35	13.91	17.08	13.73
29	119.86248	15.38053	SDSS J075926.99+152250.4	15.98	14.95	14.17	14.93	12.46
30	119.85500	15.38338	SDSS J075925.20+152259.9	16.41	15.26	14.90	14.78	14.75
31	119.84313	15.37685	SDSS J075922.30+152236.0	16.88	15.22	14.63	15.11	14.38
32	119.84477	15.36552	SDSS J075922.75+152155.6	11.83	8.80	8.01	11.07	7.60
33	119.75398	15.37014	SDSS J075900.95+152212.5	16.16	15.05	14.75	14.66	14.65
34	119.75764	15.36403	SDSS J075901.82+152150.5	15.89	14.82	14.44	14.73	14.26
35	119.71917	15.36125	SDSS J075852.58+152140.7	14.94	13.91	13.48	16.09	13.36
36	119.72440	15.35343	SDSS J075853.85+152112.3	15.64	14.47	14.13	16.01	14.00
37	119.68601	15.35741	SDSS J075844.63+152126.5	14.55	11.93	11.98	12.02	13.17
38	119.85361	15.34760	SDSS J075924.86+152051.2	16.41	15.19	14.75	14.58	14.54
39	119.79249	15.34633	SDSS J075910.19+152046.6	17.17	15.68	15.13	14.93	14.84
40	119.81588	15.33937	SDSS J075915.80+152021.4	14.89	14.99	14.76	15.27	13.02
41	119.83754	15.32322	SDSS J075921.01+151923.4	17.10	15.40	14.69	14.50	14.35
42	119.80397	15.31666	SDSS J075912.94+151859.7	14.64	14.53	14.20	11.79	13.15
43	119.79459	15.31958	SDSS J075910.70+151910.3	14.60	12.18	12.52	14.76	13.11
44	119.78983	15.30434	SDSS J075909.56+151815.7	15.47	14.23	15.74	14.19	13.63
45	119.66704	15.32577	SDSS J075840.08+151932.7	14.79	13.28	12.95	12.84	13.13
46	119.76095	15.28545	SDSS J075902.62+151707.4	16.55	14.80	14.15	15.96	13.82
47	119.76455	15.27336	SDSS J075903.49+151624.0	15.89	14.52	14.06	17.33	13.85
48	119.75449	15.27806	SDSS J075901.08+151640.7	15.63	14.94	14.34	14.94	12.82
49	119.72018	15.28046	SDSS J075852.84+151649.5	16.26	14.86	13.19	12.82	13.05

Table 3.2 Observation details for our transmission spectroscopy measurements of GJ3470b (Awiphan et al. 2016a)

Observation date	Telescope	Filter	Exposure (s)	Number of images
17 December 2013	TNO 0.5 m	Cousins-R	20.0,15.0*	584
06 January 2014	TNO 0.5 m	Cousins-R	15.0	500
10 January 2014	PROMPT-8	Cousins-R	10.0	425
04 March 2014	TNT	Sloan z'	5.65	1883
14 March 2014	TNT	Sloan r'	8.29	1777
03 April 2014	TNT	Sloan z'	5.65	1254
22 January 2015	PROMPT-8	Cousins-R	10.0	449
06 March 2015	TNT	Sloan r'	3.13	1500
16 March 2015	TNT	Sloan i'	3.13	2755
17 March 2016	TNT	Sloan g'	14.85	550

*The exposure time of the first 296 images is 20 s and the exposure time of the last 288 images is 15 s

number of frames in each observation are described below and are listed in Table 3.2. The four-minute binned light curves are shown in Fig. 3.8.

0.5 m Telescope at Thai National Observatory

Two full transit observations of GJ3470b were obtained through a Cousins-R filter using an Apogee Altra U9000 3056 × 3056 pixel CCD camera attached to the 0.5 m Schmidt–Cassegrain Telescope located at Thai National Observatory (TNO). The field-of-view of each image is 58 × 58 arcmin^2. For the first observation on 17 December 2013, the exposure time was set to be 20 s during the first half of observations (296 images) and 15 s during the second half (288 images) due to the variation in seeing at the site. On 6 January 2014, observations with a 20 s exposure time were obtained. The dead time between exposures is ∼10 s.

PROMPT 8 Telescope (0.6 m)

We observed two full transits with PROMPT 8, a 0.6 m robotic telescope at Cerro Tololo Inter-American Observatory (CTIO), Chile, with a 2048 × 2048 pixel CCD camera with a scale of 0.624 arcsec/pixel. The observations were performed with 10 s exposures and ∼20 s dead times through a Cousins-R filter on 10 January 2014 and 22 January 2015.

Thai National Telescope (2.4 m)

We conducted photometric observations of GJ3470 with ULTRASPEC (Dhillon et al. 2014), a 1k × 1k pixel high-speed frame-transfer EMCCD camera, on the 2.4 m Thai National Telescope (TNT) at TNO during the 2013-16 observing seasons (Fig. 3.5). The camera has a field of view 7.68 × 7.68 arcmin^2. The dead time between exposures is only 14 ms. Observations through the z', i', r' and g' filters were performed on separate nights (Fig. 3.6).

Fig. 3.5 Interior (left) and exterior (right) photographs of the TNT (Dhillon et al. 2014)

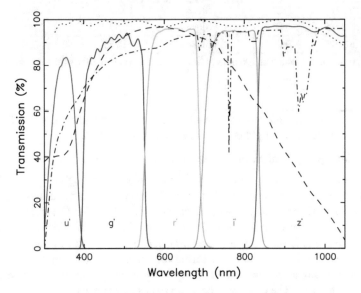

Fig. 3.6 Transmission profiles of the ULTRASPEC SDSS filter set. The transmission of one of the anti-reflection coatings used on the ULTRASPEC lenses (the CaF2 elements), the transmission of the atmosphere for unit airmass and the quantum efficiency curve of the ULTRASPEC EMCCD are shown by the dotted, dasheddotted and dashed lines, respectively (Dhillon et al. 2014)

During the 2013–2014 observing season, we were not able to perform observations with our intended filters, due to a filter wheel problem at TNT. On 4 March 2014 and 3 April 2014 observation, we aimed to observe with an i' band filter. However, a z' band filter was used instead. On 14 March 2014, our r' band light curve during poor seeing conditions was obtained, instead of a g' band light curve. We subsequently monitored GJ3470 for two nights in 2015 and one night in 2016 to obtain light curves in i', r' and g' bands. The sample image is shown in Fig. 3.7.

The host star, GJ3470, is an M-dwarf. Therefore, in Fig. 3.8, the g' band light curve provides smaller signal-to-noise ratio compared to other light curves. Although the

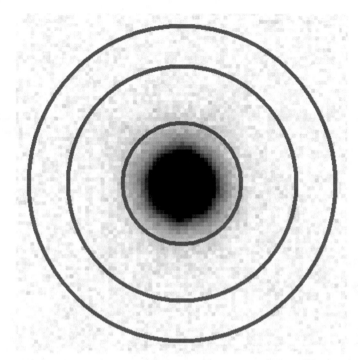

Fig. 3.7 Photometric image of GJ3470b with 5.65 s exposure time from 4 March 2014 observation with 15, 30 and 40 pixel radius aperture, inner annulus and outer annulus, respectively

g' filter light curve shows a large scatter, it is still an important inclusion for planetary atmosphere modelling, especially for Rayleigh scattering curve fitting (Sect. 3.6.1).

3.4.2 Light Curve Analysis

The calibration was carried out using the DAOPHOT package and the photometry was carried out using Python scripts which perform aperture photometry. In order to fit the light curves, we use the Transit Analysis Package (TAP, Gazak et al. 2012), a set of IDL routines which employs the Markov Chain Monte Carlo (MCMC) technique of Mandel and Agol (2002).

We combined our 10 light curves with 4 light curves from Bonfils et al. (2012), in order to fit their mid transit times. Although Biddle et al. (2014) re-analyzed and fit the mid-transit times of the Bonfils et al. (2012) data, they did not include the 12th April 2012 transit.

We set scales for semi-major axis (a/R_*), period (P) and inclination (i) to be consistent for all light curves. The planet-star radius ratio (R_p/R_*) and quadratic limb darkening coefficients are taken to be filter dependent. The mid-transit times

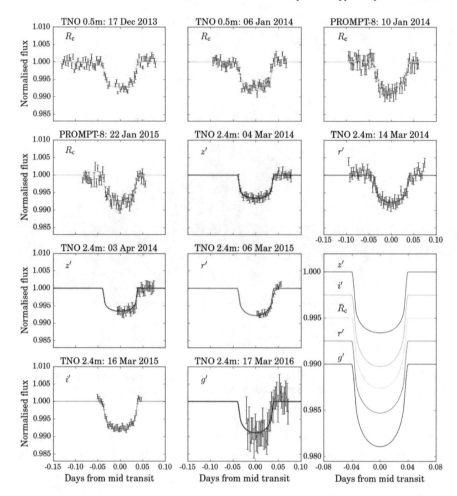

Fig. 3.8 Light curves of GJ3470b with 4 min binning and with the best fit model from the TAP analysis. The best fit model light curves in all filters are shown in the bottom right panel with arbitrary off-sets (-0.0025, -0.0050, -0.0075 and -0.0100). Thick yellow, red, orange, green and blue lines represent the best fit model in Cousins-R, z', i', r' and g' filters, respectively (Awiphan et al. 2016a)

(T_0) observed at the same epoch are also fixed to be the same value. From the previous studies, the eccentricity of the system is less than 0.051 (Bonfils et al. 2012). Therefore, in this work, a circular orbit is assumed. 1,000,000 MCMC steps are performed and the best fits from TAP are shown in Fig. 3.8.

The orbital elements calculated by TAP are compared with the results from previous studies in Table 3.3. The results from TAP provide a compatible planetary orbital period, inclination and scaled semi-major axis that agree to within 2-σ with results from previous studies. Table 3.4 compares the quadratic limb darkening coefficients (u_1 and u_2) with the values from the Claret and Bloemen (2011) catalogue, which

Table 3.3 GJ3470b orbital elements from our TAP analysis and from previous studies (Awiphan et al. 2016a)

Reference	Orbital period (Days)	Inclination (Degree)	a/R_*	R_p/R_*	u_1	u_2	Filter
Bonfils et al. (2012)	3.33714 ± 0.00017	>88.8	14.9 ± 1.2	0.0755 ± 0.0031	$0.04^{[1]}$	$0.19^{[1]}$	Gunn-Z
						$0.40^{[1]}$	No filter
Demory et al. (2013)	3.33665 ± 0.00005	$88.3^{+0.5}_{-0.4}$	$13.42^{+0.55}_{-0.53}$	$0.07798^{+0.00046}_{-0.00045}$	0.033 ± 0.015	0.181 ± 0.10	IRAC4.5 μm
Fukui et al. (2013)	3.336648 ± 0.000005	-	$14.02^{+0.33}_{-0.39}$	$0.07577^{+0.00072}_{-0.00075}$	$0.137^{+0.077}_{-0.073}$	$0.255^{[1]}$	J
				0.0802 ± 0.0013	0.19 ± 0.11	$0.338^{[1]}$	I_c
				0.0776 ± 0.0038	0.25 ± 0.15	$0.322^{[1]}$	R_c
				0.0809 ± 0.0031	$0.486^{[1]}$	$0.289^{[1]}$	g
Nascimbeni et al. (2013)	3.336649 ± 0.000002	$88.12^{+0.34}_{-0.30}$	-	$0.07484^{+0.00052}_{-0.00048}$	0.25 ± 0.04	0.49 ± 0.03	F972N20
				0.0821 ± 0.0013	0.30 ± 0.03	0.29 ± 0.03	U
Crossfield et al. (2013)	3.336665 ± 0.000002	$88.98^{+0.94}_{-1.25}$	-	$0.0789^{+0.0021}_{-0.0019}$	$-0.351^{+0.025}_{-0.023}$	$-0.889^{+0.051}_{-0.052}$	2.09-2.36 μm
Biddle et al. (2014)	$3.336648^{+0.0000043}_{-0.0000033}$	$88.88^{+0.44}_{-0.45}$	$13.94^{+0.44}_{-0.49}$	$0.0766^{+0.0019}_{-0.0020}$	$0.017^{+0.014}_{-0.012}$	0.5030 ± 0.0068	Gunn-Z
				$0.0766^{+0.0019}_{-0.0020}$	$0.029^{+0.025}_{-0.018}$	0.5030 ± 0.014	Panstarrs-Z
				$0.0765^{+0.0027}_{-0.0030}$	$0.123^{+0.038}_{-0.047}$	0.488 ± 0.020	i'
				$0.0780^{+0.0015}_{-0.0016}$	0.070 ± 0.025	$0.517^{+0.010}_{-0.009}$	I
				$0.0736^{+0.0029}_{-0.0031}$	$0.083^{+0.035}_{-0.032}$	0.519 ± 0.016	Arizona-I
				0.0803 ± 0.0025	$0.403^{+0.040}_{-0.044}$	$0.390^{+0.036}_{-0.038}$	r'
				$0.084^{+0.013}_{-0.016}$	-	-	Bessel-B
Dragomir et al. (2015)	3.3366413 ± 0.0000060	-	$12.92^{+0.72}_{-0.65}$	$0.0771^{+0.0012}_{-0.0011}$	0.123 ± 0.050	0.489 ± 0.050	$i'^{[2]}$
				$0.0770^{+0.0020}_{-0.0019}$	0.360 ± 0.050	0.411 ± 0.050	Harris-V
				0.0833 ± 0.0019	0.398 ± 0.050	0.390 ± 0.050	g'
				$0.0827^{+0.0002}_{-0.0003}$	0.421 ± 0.050	0.398 ± 0.050	Harris-B
This work	$3.3366496^{+0.0000039}_{-0.0000033}$	$89.13^{+0.26}_{-0.34}$	$13.98^{+0.20}_{-0.28}$	$0.0744^{+0.0020}_{-0.0020}$	$0.356^{+0.091}_{-0.094}$	$0.307^{+0.091}_{-0.112}$	z'
				$0.0785^{+0.0008}_{-0.0008}$	$0.469^{+0.026}_{-0.046}$	$0.350^{+0.031}_{-0.074}$	i'
				$0.0765^{+0.0017}_{-0.0015}$	$0.585^{+0.023}_{-0.054}$	$0.278^{+0.045}_{-0.091}$	R_c
				$0.0787^{+0.0016}_{-0.0022}$	$0.540^{+0.079}_{-0.047}$	$0.212^{+0.081}_{-0.077}$	r'
				$0.0832^{+0.0027}_{-0.0027}$	$0.568^{+0.062}_{-0.094}$	$0.304^{+0.068}_{-0.099}$	g'
				$0.0752^{+0.0030}_{-0.0029}$	$0.22^{+0.14}_{-0.11}$	$0.15^{+0.17}_{-0.09}$	Gunn-Z [3]
				$0.0913^{+0.0047}_{-0.0053}$	$0.45^{+0.17}_{-0.20}$	$0.33^{+0.22}_{-0.20}$	No filter [3]

Remark

[1] Fixed value

[2] Re-analyzed Biddle et al. (2014) data

[3] Re-analyzed Bonfils et al. (2012) data

Table 3.4 GJ3470 quadratic limb darkening coefficients from our TAP analysis together with predicted coefficients from the models of Claret and Bloemen (2011), with both least-square (L) and flux conservation (F) fitting methods (Awiphan et al. 2016a)

Filter	Best fit		Claret L		Claret F	
	u_1	u_2	u_1	u_2	u_1	u_2
R	$0.585^{+0.023}_{-0.054}$	$0.278^{+0.045}_{-0.091}$	0.4998	0.2329	0.5179	0.2101
g'	$0.568^{+0.062}_{-0.094}$	$0.304^{+0.068}_{-0.099}$	0.5154	0.3046	0.5405	0.2724
r'	$0.540^{+0.079}_{-0.047}$	$0.212^{+0.081}_{-0.077}$	0.5419	0.2221	0.5572	0.2028
i'	$0.469^{+0.026}_{-0.046}$	$0.350^{+0.031}_{-0.074}$	0.3782	0.2830	0.4053	0.2486
z'	$0.356^{+0.081}_{-0.094}$	$0.307^{+0.091}_{-0.112}$	0.3804	0.2361	0.2746	0.3311

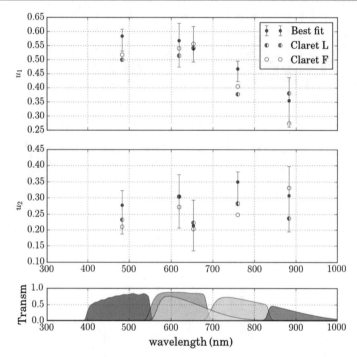

Fig. 3.9 GJ3470 quadratic limb darkening coefficients, u_1 (Top) and u_2 (Middle) from our TAP analysis (Red full-filled) together with predicted coefficients from the models of Claret and Bloemen (2011), with both least-square (L, Green half-filled) and flux conservation (F, Blue circle) fitting methods. (Bottom) The bandpass of our filters: g', r', Cousins-R, i' and z' band (from left to right). The bandpass colours have the same description as Fig. 3.8 (Awiphan et al. 2016a)

is based on the PHOENIX model,[1] with both least-square and flux conservation fitting methods. We use the limb darkening coefficients of a star with stellar temperature $T_{\rm eff} = 3500$ K, surface gravity $\log(g_*) = 4.5$ and metallicity [Fe/H] = 0.2, which is the nearest grid point ($T_{\rm eff} = 3600 \pm 100$ K, $\log(g_*) = 4.658 \pm 0.035$ and [Fe/H] = 0.20 ± 0.10 (Demory et al. 2013)). The Claret and Bloemen (2011) catalogue provides compatible (within 2-σ variation) limb darkening coefficients with the best fit coefficients from TAP (Fig. 3.9). Therefore, in following work, we use best fit limb darkening coefficients from TAP to obtain planetary parameters.

3.4.3 Stellar and Planetary Characterizations

In order to obtain planetary physical parameters, the parameters of the host star must also be considered. The mean stellar density is calculated by Kepler's third law neglecting planetary mass. From the TAP result, the mean density of GJ3470 is $\rho_* = 3.30 \pm 0.17 \rho_\odot$. This result is consistent with the value derived by other works (Table 3.5).

To find other stellar and planetary parameters, we adopt a stellar mass, $M_* = 0.539^{+0.047}_{-0.043} M_\odot$ from Demory et al. (2013), which was obtained from the average of the $J-$, $H-$, and $K-$band mass-luminosity (M-L) relations of Delfosse et al. (2000). We also adopt a radial velocity amplitude parameter $K' = 13.4 \pm 1.2$ m s^{-1}d$^{1/3}$ from Demory et al. (2013), where

$$K' = K P_p^{1/3} = \frac{(2\pi G)^{1/3} M_p \sin i}{(M_* + M_p)^{2/3}} \tag{3.33}$$

for a circular orbit. In the above equation, K is the radial velocity semi-amplitude, M_p is the planet mass, M_* is the host star mass and G is the gravitational constant.

Combining with the mean density, the calculated radius of the GJ3470 host star is $R_* = 0.547 \pm 0.018 R_\odot$. The planetary radius is calculated from the planet-star radius ratio, which is wavelength dependent. We use the ratio in the Cousins-R waveband to calculate a radius, $R_p = 4.57 \pm 0.18 R_\oplus$.

The calculated planetary mass and density are $M_p = 13.9 \pm 1.5 M_\oplus$ and $\rho_p = 0.80 \pm 0.13$ g cm^{-3}. The range of planetary equilibrium temperature, T_p, can be derived from the relation,

$$T_p = T_{\text{eff}} \left(\frac{1-A}{4F} \right)^{1/4} \left(\frac{R_*}{2a_p} \right)^{1/2} \tag{3.34}$$

Table 3.5 Mean stellar density of GJ3470 from our TAP analysis and previous studies (Awiphan et al. 2016a)

Reference	Stellar density ($\rho_\odot$)
Bonfils et al. (2012)	4.26 ± 0.53
Demory et al. (2013)	2.91 ± 0.37
Pineda et al. (2013)	4.25 ± 0.40
Fukui et al. (2013)	3.32 ± 0.27
Nascimbeni et al. (2013)	2.74 ± 0.19
Crossfield et al. (2013)	3.49 ± 1.13
Biddle et al. (2014)	$3.39^{+0.30}_{-0.32}$
This work	3.30 ± 0.17

---

¹ See http://phoenix.ens.lyon.fr/simulator/

Table 3.6 Summary of GJ3470b properties (Awiphan et al. 2016a)

Parameter	Symbol	Value	Unit
Stellar parameters			
Stellar mass	M_*	$0.539^{+0.047}_{-0.043}$ *	$M_\odot$
Stellar radius	R_*	0.547 ± 0.018	$R_\odot$
Stellar density	ρ_*	3.30 ± 0.17	g cm^{-3}
Stellar surface gravity	$\log(g_*)$	4.695 ± 0.046	cgs
Stellar effective temperature	T_{eff}	3600 ± 100*	K
Stellar metallicity	[Fe/H]	0.20 ± 0.10*	
Planetary parameters			
Orbital period	P	$3.3366496^{+0.0000039}_{-0.0000033}$	d
Orbital inclination	i	$89.13^{+0.26}_{-0.34}$	deg
Semi-major axis	a	0.0355 ± 0.0019	AU
Epoch of mid-transit (BJD)	T_0	2455983.70421	d
Radial velocity amplitude parameter	K'	13.4 ± 1.2 *	m s^{-1}d$^{1/3}$
Planetary mass	M_p	13.9 ± 1.5	$M_\oplus$
Planetary radius	R_p	4.57 ± 0.18	$R_\oplus$
Planetary density	ρ_p	0.80 ± 0.13	g cm^{-3}
Planetary equilibrium temperature	T_p	497–690	K
Planetary surface gravity	$\log(g_p)$	2.815 ± 0.057	cgs
Planetary atmospheric scale height	h	760 ± 140	km
Planetary atmospheric mean molecular weight	μ	1.08 ± 0.20	

*Adopted value from Demory et al. (2013)

In the calculation, we use $T_{\mathrm{eff}} = 3600$ K and $a/R_* = 13.98$ and their uncertainties are not taken into account. The temperature range in our work, due to the possible range of Bond albedo and the heat redistribution factor, is $T_p = 497$–690 K. The list of all parameters from the analysis is shown in Table 3.6.

3.5   Transit Timing Variations Analysis of GJ3470b

3.5.1   O-C Diagram

The measured mid-transit time of a photometric light curve always has some variation due to noise. However, as discussed in Sect. 3.1.5, variations can be caused by the gravitational interaction of other objects in the system, such as other exoplanets or exomoons. Therefore, we can use the mid-transit times of GJ3470b to place limits on transit timing variations. From the TAP analysis, the mid-transit times of 14 light curves are shown in Table 3.7.

Table 3.7 Mid transit of GJ3470b from our TAP analysis (Awiphan et al. 2016a)

Observing date	Mid-transit time (BJD) (BJD-2450000)	$(O - C)$ (d)
26 February 2012*	$5983.7015^{+0.0015}_{-0.0015}$	-0.0026
07 March 2012*	$5993.7152^{+0.0014}_{-0.0014}$	0.0012
12 April 2012*	$6030.4177^{+0.0010}_{-0.0010}$	0.0006
17 December 2013	$6644.3602^{+0.0011}_{-0.0014}$	-0.0006
06 January 2014	$6664.3812^{+0.0017}_{-0.0017}$	0.0005
10 January 2014	$6667.7169^{+0.0021}_{-0.0021}$	-0.0004
04 March 2014	$6721.1038^{+0.0004}_{-0.0005}$	0.0001
14 March 2014	$6731.1162^{+0.0011}_{-0.0011}$	0.0026
03 April 2014	$6751.1332^{+0.0007}_{-0.0008}$	-0.0004
22 January 2015	$7044.7603^{+0.0019}_{-0.0024}$	0.0015
06 March 2015	$7088.1370^{+0.0008}_{-0.0007}$	0.0018
16 March 2015	$7098.1455^{+0.0004}_{-0.0005}$	0.0003
17 March 2016	$7465.1750^{+0.0017}_{-0.0014}$	-0.0017

*Re-analyzed Bonfils et al. (2012) data

Fig. 3.10 O-C diagram of exoplanet GJ3470b. Epoch = 0 is the transit on 26 February 2012. The red filled, cyan half-filled and unfilled markers represent the mid-transit time from our observations, re-analysed Bonfils et al. (2012) observations and other previous published observations, respectively (Awiphan et al. 2016a)

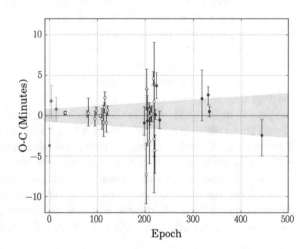

We use these mid-transit times, and those from previous studies, to plot the epoch of each transit against the observed minus the calculated time $(O - C)$ in order to find the TTV of GJ3470b. We perform a linear fit to the $O - C$ diagram to correct GJ3470b's ephemeris. The best linear fit (reduced chi-squared, $\chi^2_{r,L} = 2.11$) gives a corrected ephemeris of

$$T_0(E) = 2455983.70421(\pm 0.00053) + 3.33665(\pm 0.00001)E \quad \text{(BJD)}, \quad (3.35)$$

where E is the number of epochs from the 26 February 2012 transit, the first transit of Bonfils et al. (2012). In Fig. 3.10, the $O - C$ diagram shows that there is no significant variation of the mid-transit time. Almost all of them are consistent within 2-σ.

3.5.2 Upper Mass Limit of the Second Planet

From Sect. 3.5.1, the $O - C$ diagram shows no significant TTV signal, indicating that there are no nearby massive objects, which have strong gravitational interaction with GJ3470b. We use this to compute an upper mass limit on a second planet in the system. We assume that the second planet is in a circular orbit that is also coplanar with GJ3470b's orbit. We use the `TTVFaster` code (Agol and Deck 2016), which computes the TTV signal from analytic formulae.

We employ two methods to measure the upper mass limit. First, we calculate the TTV signal for the second planet over a mass range from $10^{-1} M_\oplus$ to $10^3 M_\oplus$ with steps of 0.01 in $\log(M)$. We sample a period ratio of the perturbing planet and GJ3470b over a ratio range from 0.30 to 4.50 with 0.01 steps. At each grid point, the initial phase of the perturber is varied between 0 and 2π with $\pi/18,000$ steps in order to cover all alignments of the second planet at $E = 0$. For each period of the perturber planet, the minimum mass which produces a TTV signal higher than the measured TTV limit is taken to be the upper mass limit for that period. As the highest TTV signal from the $O - C$ diagram is 498 s, upper mass limits corresponding to TTV amplitudes of 400, 500 and 600 s are calculated and shown in Fig. 3.11.

The second method uses the reduced chi-squared of the best-fit between the observed TTV signal and the signal from `TTVFaster`. The grid points and the initial phase of second planet are varied as in the first method. From Sect. 3.5.1, the best linear fit using a single-planet model is $\chi^2_{r,L} = 2.11$. We assess the improvement to the fit of introducing a second planet though the delta reduced chi-squared statistic, $\Delta\chi^2_r = \chi^2_r - \chi^2_{r,L}$, where χ^2_r is the best fitting TTV model at the given mass and period. In Fig. 3.11, $\Delta\chi^2_r$ is shown as a function of perturber mass and period. The preferred planet models are shown in Fig. 3.11 as negative valued $\Delta\chi^2_r$ regions. The best fit TTV models are shown with the black dotted line in Fig. 3.11, which is produced by averaging over period ratio bins of width 0.05.

Unstable orbit regions are calculated from the mutual Hill sphere between GJ3470b and the perturber. For two-planet systems in coplanar and circular orbits, the boundary of the stable orbit is when the separation of the planets' semi-major axes ($a_{\text{out}} - a_{\text{in}}$) is larger than $2\sqrt{3}$ of the mutual Hill sphere (Fabrycky et al. 2012)

$$r_H = \frac{a_{\text{in}} + a_{\text{out}}}{2} \left(\frac{M_{\text{in}} + M_{\text{out}}}{3M_*} \right)^{1/3}. \tag{3.36}$$

In Eq. 3.36, a_{in} and a_{out} are the semi-major axis of the inner and outer planets, respectively. The area of unstable orbits is shown by the black shaded region of Fig. 3.11. Orbital resonances between GJ3470b and the second planet are shown as vertical lines. In the cases where GJ3470b and the perturber are in a first-order mean motion resonance, the upper mass limits are lower.

From Fig. 3.11, the TTV area, with a $\Delta\chi^2_r$ between -0.7 and -0.4 with 35 degrees of freedom is shown near the upper mass-limit of 400 s TTV amplitude. A nearby second planet with period between 2.5 and 4.0 days is ruled out by both upper-mass

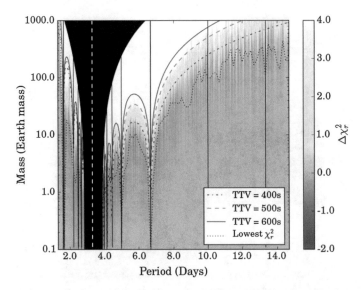

Fig. 3.11 Upper mass limit of a second planet in the GJ3470 system. The blue dashed-dot, green dashed and red solid lines represent the upper mass limit for 400, 500 and 600 s TTV amplitudes. The contours show the $\Delta\chi_r^2$ between the best TTV fit and the best linear fit. The black dotted line presents the best $\Delta\chi_r^2$ within a 0.05 period ratio bin. From left to right, the black vertical lines show 3:1, 2:1, 3:2, 4:3, 5:6, 4:5, 3:4, 2:3, 1:2, 1:3 and 1:4 resonance periods. The white vertical dashed line shows the orbital period of GJ3470b (Awiphan et al. 2016a)

limit tests and the mutual Hill sphere area. A Jupiter-mass planet with period less than 10 days is also excluded. From this result, we can conclude that there is no nearby massive planet to GJ3470b.

3.6   Transmission Spectroscopy Analysis of GJ3470b

3.6.1   Rayleigh Scattering

As discussed in Sect. 3.1.6, transmission spectroscopy of an exoplanet can be seen as a change in planet-star radius ratio as a function of wavelength. From Eq. 3.25, we adopt a Bond Albedo, $A = 0.3$ and an equilibrium temperature, $T = 624 \pm 25$ K from Demory et al. (2013). We assume the main physical process involved in GJ3470b's atmosphere is Rayleigh scattering ($\alpha = -4$) without atomic or molecular absorption. In order to find $dR_p/d\ln\lambda$, we combine our planet-star radius ratio with previous optical observations. The plot of planet-star radius ratio versus wavelength and the best fit model of the GJ3470b data with mean-molecular weight 1.00, 1.50, 2.22 (Jupiter) and 2.61 (Neptune) atmospheres are shown in Fig. 3.12.

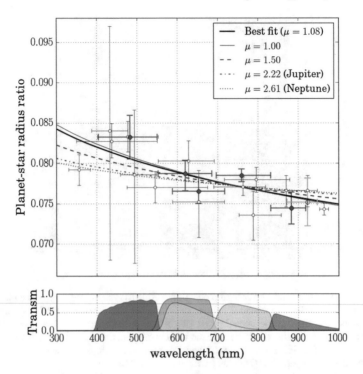

Fig. 3.12 (Top) Relation between planet-star radius ratio and wavelength in optical filters with $dR_p/d\ln\lambda$ slope fitting (Black thick line). The blue solid, dashed, dashed-dot and dotted lines represent the best fit with mean-molecular weight 1.00, 1.50, 2.22 (Jupiter) and 2.61 (Neptune), respectively. The markers have the same description as Fig. 3.10. (Bottom) The bandpass of our filters: g', r', Cousins-R, i' and z' band (from left to right). The bandpass colours have the same description as Fig. 3.8 (Awiphan et al. 2016a)

From the curve fit to the data in Fig. 3.12 and Eq. 3.25, a low mean molecular weight of 1.08 ± 0.20 is obtained. This low mean molecular weight is consistent with a H/He-dominated atmosphere as in previous studies (Nascimbeni et al. 2013: $1.32^{+0.27}_{-0.19}$ and Dragomir et al. 2015: 1.35 ± 0.44).

3.6.2 Atmospheric Composition

To determine the atmospheric composition, ideally a detailed atmospheric model of GJ3470b is required. We rescale the planetary atmosphere models of Howe and Burrows (2012). We use 114 models which have planetary equilibrium temperature and mass close to GJ3470b ($T_p = 700$ K, $M_p = 10 M_\oplus$) as listed in Table 3.8. The optical planet-star radius ratio data from Sect. 3.6.1 and infrared data from Demory et al. (2013), Ehrenreich et al. (2014) are used to compute a reduced chi-squared

($\chi^2_{r,\text{atm}}$) between the rescaled models and the data. In Table 3.8, the 114 atmosphere models with their best fit $\chi^2_{r,\text{atm}}$ are shown (Fig. 3.13).

From the fitting, a CH_4 atmosphere with a $100\,\text{cm}^{-3}$ particle abundance haze at $1-1000\,\mu\text{bar}$ altitude provides the best fit with $\chi^2_{r,\text{atm}} = 1.40$ or 1.38 for polyacetylene or tholin haze, respectively. At near-infrared wavelengths, CH_4 models provide the best fit and the data are not compatible with a cloudy atmosphere model. However, the models of Howe and Burrows (2012) do not provide an atmosphere with mixed composition (the CH_4 atmosphere is a 100% methane atmosphere), which might be the cause of the poor fit to the data at optical wavelengths. From the Rayleigh scattering slope, the mean molecular weight is too low to be a methane dominated atmosphere. Therefore, the H/He dominated haze with high particle abundance, such as high altitude polyacetylene and tholin with a methane contaminant, is preferred. A model atmosphere with a mixed-ratio composition should provide a better description of the GJ3470b atmosphere.

3.7 Conclusion

In this Chapter, we have studied a transiting hot Neptune, GJ3470b, which is the first sub-Jovian planet with detected Rayleigh scattering. Optical multi-filter observations of the exoplanet were obtained with the 2.4-m and 0.5-m telescopes at the Thai National Observatory (TNO) and the 0.6-m telescope at Cerro Tololo Inter-American Observatory (CTIO) in 2013–2016. Ten transit light curves were obtained and analyzed using the TAP program (Agol et al. 2005). From the analysis, we obtain a planet mass $M_p = 13.9 \pm 1.5 M_\oplus$, radius $R_p = 4.74 \pm 0.32 R_\oplus$, period $P = 3.3366496^{+0.0000039}_{-0.0000033}$ d, and inclination $i = 89.13^{+0.26}_{-0.34}$ degrees. A new ephemeris for GJ3470b is also provided.

We perform the TTV analysis with the TTVFaster code of Agol and Deck (2016), in order to determine an upper mass limit for a second planet in the system. The TTV signal indicates little variation, which excludes the presence of a hot-Jupiter with orbital period less than 10 d in the system. The mutual Hill sphere also excludes the presence of a nearby planet with orbital period between 2.5 and 4.0 d.

For the transmission spectroscopy analysis, GJ3470b's low atmosphere mean molecular weight ($\mu = 1.08 \pm 0.20$) is obtained from the Rayleigh scattering fitting of the planet-star radius ratio variation in the optical. We confirm the steep Rayleigh scattering slope favoured by previous studies. Previous near-infrared data favours a methane atmosphere with high particle abundance ($100\,\text{cm}^{-3}$ of tholin or polyacetylene) at high altitude ($1000 - 1\,\mu\text{bar}$) when compared to the model atmosphere of Howe and Burrows (2012). However, the models do not fit the data at optical wavelengths, which might be a consequence of the single atmosphere composition within the models. A mixed-ratio composition model could provide a better understanding of the planet's atmosphere.

Table 3.8 Atmospheric models of Howe and Burrows (2012) with the best $\chi^2_{r,\text{atm}}$ fit (Awiphan et al. 2016a)

Particle	Composition	Cloud top pressure	χ^2_r
No Cloud	$0.3\times$ solar	1 bar	3.24
No Cloud	$1\times$ solar	1 bar	4.05
No Cloud	$3\times$ solar	1 bar	4.64
No Cloud	**CH_4**	**1 bar**	**1.49**
No Cloud	CO_2	1 bar	1.79
No Cloud	H_2O	1 bar	1.86
Cloud	$0.3\times$ solar	1 mbar	1.80
Cloud	$1\times$ solar	1 mbar	2.04
Cloud	$3\times$ solar	1 mbar	2.33
Cloud	**CH_4**	**1 mbar**	**1.64**
Cloud	CO_2	1 mbar	1.76
Cloud	H_2O	1 mbar	1.77
Cloud	$0.3\times$ solar	1 μbar	1.71
Cloud	$1\times$ solar	1 μbar	1.71
Cloud	$3\times$ solar	1 μbar	1.71
Cloud	CH_4	1 μbar	1.71
Cloud	CO_2	1 μbar	1.72
0.1 μm 100 cm^{-3} polyacetylene	$0.3\times$ solar	1 μbar	2.14
0.1 μm 100 cm^{-3} polyacetylene	$1\times$ solar	1 μbar	2.36
0.1 μm 100 cm^{-3} polyacetylene	$3\times$ solar	1 μbar	2.51
0.1 μm 100 cm^{-3} polyacetylene	**CH_4**	**1 μbar**	**1.40**
0.1 μm 100 cm^{-3} polyacetylene	CO_2	1 μbar	1.78
0.1 μm 100 cm^{-3} polyacetylene	H_2O	1 μbar	1.76
0.1 μm 100 cm^{-3} polyacetylene	$0.3\times$ solar	1 mbar	2.26
0.1 μm 100 cm^{-3} polyacetylene	$1\times$ solar	1 mbar	2.79
0.1 μm 100 cm^{-3} polyacetylene	$3\times$ solar	1 mbar	3.19
0.1 μm 100 cm^{-3} polyacetylene	CH_4	1 mbar	1.47
0.1 μm 100 cm^{-3} polyacetylene	CO_2	1 mbar	1.79
0.1 μm 100 cm^{-3} polyacetylene	H_2O	1 mbar	1.82
0.1 μm 1000 cm^{-3} polyacetylene	$0.3\times$ solar	1 μbar	3.85
0.1 μm 1000 cm^{-3} polyacetylene	$1\times$ solar	1 μbar	3.41
0.1 μm 1000 cm^{-3} polyacetylene	$3\times$ solar	1 μbar	3.13
0.1 μm 1000 cm^{-3} polyacetylene	CH_4	1 μbar	1.49
0.1 μm 1000 cm^{-3} polyacetylene	CO_2	1 μbar	1.77
0.1 μm 1000 cm^{-3} polyacetylene	H_2O	1 μbar	1.74
0.1 μm 1000 cm^{-3} polyacetylene	$0.3\times$ solar	1 mbar	2.26
0.1 μm 1000 cm^{-3} polyacetylene	$1\times$ solar	1 mbar	2.50
0.1 μm 1000 cm^{-3} polyacetylene	$3\times$ solar	1 mbar	2.72
0.1 μm 1000 cm^{-3} polyacetylene	CH_4	1 mbar	1.52

Table 3.8 (continued)

Particle	Composition	Cloud top pressure	χ_r^2
0.1 μm 1000 cm^{-3} polyacetylene	CO_2	1 mbar	1.79
0.1 μm 1000 cm^{-3} polyacetylene	H_2O	1 mbar	1.81
1.0 μm 0.01 cm^{-3} polyacetylene	0.3× solar	1 μbar	1.60
1.0 μm 0.01 cm^{-3} polyacetylene	1× solar	1 μbar	1.92
1.0 μm 0.01 cm^{-3} polyacetylene	3× solar	1 μbar	2.50
1.0 μm 0.01 cm^{-3} polyacetylene	CH_4	1 μbar	1.47
1.0 μm 0.01 cm^{-3} polyacetylene	CO_2	1 μbar	1.78
1.0 μm 0.01 cm^{-3} polyacetylene	H_2O	1 μbar	1.90
1.0 μm 0.01 cm^{-3} polyacetylene	0.3× solar	1 mbar	2.04
1.0 μm 0.01 cm^{-3} polyacetylene	1× solar	1 mbar	2.60
1.0 μm 0.01 cm^{-3} polyacetylene	3× solar	1 mbar	3.10
1.0 μm 0.01 cm^{-3} polyacetylene	CH_4	1 mbar	1.51
1.0 μm 0.01 cm^{-3} polyacetylene	CO_2	1 mbar	1.79
1.0 μm 0.01 cm^{-3} polyacetylene	H_2O	1 mbar	1.85
1.0 μm 0.1 cm^{-3} polyacetylene	0.3× solar	1 μbar	2.77
1.0 μm 0.1 cm^{-3} polyacetylene	1× solar	1 μbar	3.48
1.0 μm 0.1 cm^{-3} polyacetylene	3× solar	1 μbar	4.02
1.0 μm 0.1 cm^{-3} polyacetylene	CH_4	1 μbar	1.50
1.0 μm 0.1 cm^{-3} polyacetylene	CO_2	1 μbar	1.79
1.0 μm 0.1 cm^{-3} polyacetylene	H_2O	1 μbar	1.85
1.0 μm 0.1 cm^{-3} polyacetylene	0.3× solar	1 mbar	3.00
1.0 μm 0.1 cm^{-3} polyacetylene	1× solar	1 mbar	3.78
1.0 μm 0.1 cm^{-3} polyacetylene	3× solar	1 mbar	4.36
1.0 μm 0.1 cm^{-3} polyacetylene	CH_4	1 mbar	1.50
1.0 μm 0.1 cm^{-3} polyacetylene	CO_2	1 mbar	1.79
1.0 μm 0.1 cm^{-3} polyacetylene	H_2O	1 mbar	1.86
0.1 μm 100 cm^{-3} tholin	0.3× solar	1 μbar	2.13
0.1 μm 100 cm^{-3} tholin	1× solar	1 μbar	2.31
0.1 μm 100 cm^{-3} tholin	3× solar	1 μbar	2.44
0.1 μm 100 cm^{-3} tholin	**CH_4**	**1 μbar**	**1.38**
0.1 μm 100 cm^{-3} tholin	CO_2	1 μbar	1.78
0.1 μm 100 cm^{-3} tholin	H_2O	1 μbar	1.74
0.1 μm 100 cm^{-3} tholin	0.3× solar	1 mbar	2.21
0.1 μm 100 cm^{-3} tholin	1× solar	1 mbar	2.73
0.1 μm 100 cm^{-3} tholin	3× solar	1 mbar	3.13
0.1 μm 100 cm^{-3} tholin	CH_4	1 mbar	1.46
0.1 μm 100 cm^{-3} tholin	CO_2	1 mbar	1.79
0.1 μm 100 cm^{-3} tholin	H_2O	1 mbar	1.82
0.1 μm 1000 cm^{-3} tholin	0.3× solar	1 μbar	3.61
0.1 μm 1000 cm^{-3} tholin	1× solar	1 μbar	3.24

Table 3.8 (continued)

Particle	Composition	Cloud top pressure	χ_r^2
0.1 μm 1000 cm^{-3} tholin	3$\times$ solar	1 μbar	2.99
0.1 μm 1000 cm^{-3} tholin	CH$_4$	1 μbar	1.48
0.1 μm 1000 cm^{-3} tholin	CO$_2$	1 μbar	1.67
0.1 μm 1000 cm^{-3} tholin	H$_2$O	1 μbar	1.73
0.1 μm 1000 cm^{-3} tholin	0.3$\times$ solar	1 mbar	2.22
0.1 μm 1000 cm^{-3} tholin	1$\times$ solar	1 mbar	2.46
0.1 μm 1000 cm^{-3} tholin	3$\times$ solar	1 mbar	2.70
0.1 μm 1000 cm^{-3} tholin	CH$_4$	1 mbar	1.51
0.1 μm 1000 cm^{-3} tholin	CO$_2$	1 mbar	1.68
0.1 μm 1000 cm^{-3} tholin	H$_2$O	1 mbar	1.81
1.0 μm 0.01 cm^{-3} tholin	0.3$\times$ solar	1 μbar	2.79
1.0 μm 0.01 cm^{-3} tholin	1$\times$ solar	1 μbar	3.50
1.0 μm 0.01 cm^{-3} tholin	3$\times$ solar	1 μbar	4.04
1.0 μm 0.01 cm^{-3} tholin	CH$_4$	1 μbar	1.50
1.0 μm 0.01 cm^{-3} tholin	CO$_2$	1 μbar	1.79
1.0 μm 0.01 cm^{-3} tholin	H$_2$O	1 μbar	1.85
1.0 μm 0.01 cm^{-3} tholin	0.3$\times$ solar	1 mbar	3.00
1.0 μm 0.01 cm^{-3} tholin	1$\times$ solar	1 mbar	3.79
1.0 μm 0.01 cm^{-3} tholin	3$\times$ solar	1 mbar	4.37
1.0 μm 0.01 cm^{-3} tholin	CH$_4$	1 mbar	1.50
1.0 μm 0.01 cm^{-3} tholin	CO$_2$	1 mbar	1.79
1.0 μm 0.01 cm^{-3} tholin	H$_2$O	1 mbar	1.86
1.0 μm 0.1 cm^{-3} tholin	0.3$\times$ solar	1 μbar	1.55
1.0 μm 0.1 cm^{-3} tholin	1$\times$ solar	1 μbar	1.75
1.0 μm 0.1 cm^{-3} tholin	3$\times$ solar	1 μbar	1.96
1.0 μm 0.1 cm^{-3} tholin	CH$_4$	1 μbar	1.51
1.0 μm 0.1 cm^{-3} tholin	CO$_2$	1 μbar	1.79
1.0 μm 0.1 cm^{-3} tholin	H$_2$O	1 μbar	1.82
1.0 μm 0.1 cm^{-3} tholin	0.3$\times$ solar	1 mbar	2.07
1.0 μm 0.1 cm^{-3} tholin	1$\times$ solar	1 mbar	2.63
1.0 μm 0.1 cm^{-3} tholin	3$\times$ solar	1 mbar	3.13
1.0 μm 0.1 cm^{-3} tholin	CH$_4$	1 mbar	1.51
1.0 μm 0.1 cm^{-3} tholin	CO$_2$	1 mbar	1.79
1.0 μm 0.1 cm^{-3} tholin	H$_2$O	1 mbar	1.85

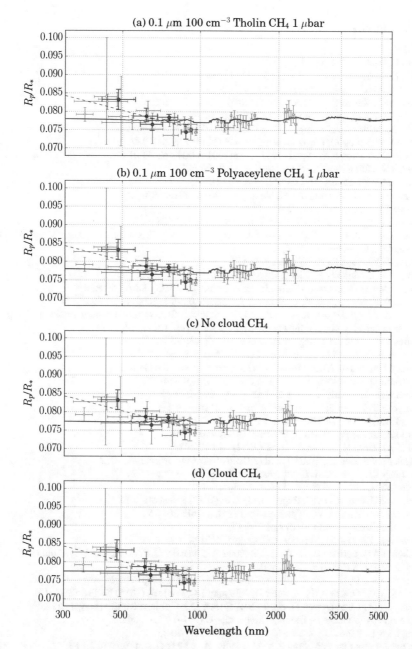

Fig. 3.13 GJ3470b atmosphere models: **a** 0.1 μm 100 cm^{-3} Tholin with 1–1000 μbar CH$_4$ ($\chi^2_{r,\text{atm}} = 1.38$), **b** 0.1 μm 100 cm^{-3} Polyacetylene with 1-1000 μbar CH$_4$ ($\chi^2_{r,\text{atm}} = 1.40$), **c** No cloud CH$_4$ ($\chi^2_{r,\text{atm}} = 1.49$) and **d** Cloud CH$_4$ ($\chi^2_{r,\text{atm}} = 1.71$), with their best $\chi^2_{r,\text{atm}}$ fits (Blue solid lines). The thin dashed lines show the Rayleigh scattering slope fitting at optical wavelength (300–1000 nm). The markers have the same description as Fig. 3.10 (Awiphan et al. 2016a)

References

Agol E, Deck K (2016) ApJ 818:177
Agol E, Steffen J, Sari R, Clarkson W (2005) MNRAS 359:567
Ahn CP, Alexandroff R, Allende Prieto C et al (2014) ApJS 211:17
Awiphan S, Kerins E, Pichadee S et al (2016a) MNRAS 463:2574
Ballard S, Fabrycky D, Fressin F et al (2011) ApJ 743:200
Bean JL, Miller-Ricci Kempton E, Homeier D (2010a) Nature 468:669
Bento J, Wheatley PJ, Copperwheat CM et al (2014) MNRAS 437:1511
Biddle LI, Pearson KA, Crossfield IJM et al (2014) MNRAS 443:1810
Bonfils X, Gillon M, Udry S et al (2012) A&A 546:A27
Borucki WJ, Summers AL (1984) Icarus 58:121
Boué G, Oshagh M, Montalto M, Santos NC (2012) MNRAS 422:57
Brown TM (2001) ApJ 553:1006
Carter JA, Agol E, Chaplin WJ et al (2012) Science 337:556
Charbonneau D, Brown TM, Burrows A, Laughlin G (2007) Protostars and planets V, pp 701–716
Claret A (2000) A&A 363:1081
Claret A, Bloemen S (2011) A&A 529:A75
Coughlin JL, Mullally F, Thompson SE et al (2016) ApJS 224:12
Crossfield IJM, Barman T, Hansen BMS, Howard AW (2013) A&A 559:A33
Czesla S, Huber KF, Wolter U, Schröter S, Schmitt JHMM (2009) A&A 505:1277
Da Costa GS (1992) In: Howell SB (ed) Astronomical CCD observing and reduction techniques.
    Astronomical Society of the Pacific Conference Series, vol 23, p 90
Delfosse X, Forveille T, Ségransan D et al (2000) A&A 364:217
Demory B-O, Torres G, Neves V et al (2013) ApJ 768:154
Dhillon VS, Marsh TR, Atkinson DC et al (2014) MNRAS 444:4009
Dragomir D, Benneke B, Pearson KA et al (2015) ApJ 814:102
Ehrenreich D, Bonfils X, Lovis C et al (2014) A&A 570:A89
Fabrycky DC, Ford EB, Steffen JH et al (2012) ApJ 750:114
Ford EB, Fabrycky DC, Steffen JH et al (2012a) ApJ 750:113
Ford EB, Ragozzine D, Rowe JF et al (2012b) ApJ 756:185
Fukui A, Narita N, Kurosaki K et al (2013) ApJ 770:95
Gazak JZ, Johnson JA, Tonry J et al (2012) Advances in astronomy, 2012
Grillmair CJ, Burrows A, Charbonneau D et al (2008) Nature 456:767
Holman MJ, Murray NW (2005) Science 307:1288
Holman MJ, Fabrycky DC, Ragozzine D et al (2010) Science 330:51
Howard AW, Marcy GW, Bryson ST et al (2012) ApJS 201:15
Howe AR, Burrows AS (2012) ApJ 756:176
Howe AR, Burrows A, Verne W (2014) ApJ 787:173
Hubbard WB, Fortney JJ, Lunine JI, Burrows A, Sudarsky D, Pinto P (2001) ApJ 560:413
Hui L, Seager S (2002) ApJ 572:540
Jontof-Hutter D, Lissauer JJ, Rowe JF, Fabrycky DC (2014) ApJ 785:15
Jontof-Hutter D, Rowe JF, Lissauer JJ, Fabrycky DC, Ford EB (2015) Nature 522:321
Knutson HA, Charbonneau D, Noyes RW, Brown TM, Gilliland RL (2007) ApJ 655:564
Knutson HA, Benneke B, Deming D, Homeier D (2014) Nature 505:66
Kreidberg L, Bean JL, Désert J-M et al (2014) Nature 505:69
Lecavelier Des Etangs A, Pont F, Vidal-Madjar A, Sing D (2008) A&A 481:L83
Lissauer JJ, Fabrycky DC, Ford EB et al (2011) Nature 470:53
Lissauer JJ, Jontof-Hutter D, Rowe JF et al (2013) ApJ 770:131
Lopez ED, Fortney JJ (2014) ApJ 792:1
Mandel K, Agol E (2002) ApJ 580:L171
Marcy GW, Weiss LM, Petigura EA, Isaacson H, Howard AW, Buchhave LA (2014) Proc Natl Acad
    Sci 111:12655

Masuda K (2014) ApJ 783:53
Mazeh T, Nachmani G, Holczer T et al (2013) ApJS 208:16
Miller-Ricci E, Fortney JJ (2010) ApJ 716:L74
Morton TD, Bryson ST, Coughlin JL et al (2016) ApJ 822:86
Nascimbeni V, Piotto G, Pagano I, Scandariato G, Sani E, Fumana M (2013) A&A 559:A32
Nesvorný D, Morbidelli A (2008) ApJ 688:636
Perryman M, Hainaut O, Dravins D et al (2005). arXiv:0506.163
Pineda JS, Bottom M, Johnson JA (2013) ApJ 767:28
Piso A-MA, Youdin AN, Murray-Clay RA (2015) ApJ 800:82
Rafikov RR (2011) ApJ 727:86
Seager S, Deming D (2010) ARA&A 48:631
Seager S, Mallén-Ornelas G (2003) ApJ 585:1038
Seager S, Sasselov DD (2000) ApJ 537:916
Silva AVR (2003) ApJ 585:L147
Southworth J, Wheatley PJ, Sams G (2007) MNRAS 379:L11
Steffen JH, Fabrycky DC, Ford EB et al (2012a) MNRAS 421:2342
Steffen JH, Ford EB, Rowe JF et al (2012b) ApJ 756:186
Steffen JH, Fabrycky DC, Agol E et al (2013) MNRAS 428:1077
Swain MR, Vasisht G, Tinetti G (2008) Nature 452:329
Swain MR, Deroo P, Griffith CA et al (2010) Nature 463:637
Tingley B, Sackett PD (2005) ApJ 627:1011
Veras D, Ford EB, Payne MJ (2011) ApJ 727:74
Weiss LM, Marcy GW (2014) ApJ 783:L6
Winn JN (2010). arXiv:1001.2010

Chapter 4
Detectability of Habitable Exomoons

Exomoons are natural satellites of exoplanets. To date, none has been confirmed yet, nevertheless two possible exomoon candidates: an exomoon orbiting a gas giant, MOA-2011-BLG-262L b (Bennett et al. 2014) and a Neptune-sized exomoon orbiting a Jupiter-sized exoplanet, Kepler-1625b I (Teachey et al. 2018) have been announced. However, if our Solar System is typical, then exomoons must be common. Several methods have been developed to detect exomoons; including the transit method (Simon et al. 2007), microlensing (Han and Han 2002; Liebig and Wambsganss 2010), pulsar-timing (Lewis et al. 2008), Rossiter–McLaughlin effect (Simon et al. 2010) and scatter-peak (Simon et al. 2012). Extensions of the transit method involving transit timing, called transit timing variation (TTV) and transit duration variation (TDV), appear to offer the best potential to detect habitable exomoons in the near future (Kipping 2011b).

The detectability of habitable exomoons orbiting giant planets in M-dwarf systems using the correlation between transit timing variation (TTV) and transit timing duration (TDV) signals with *Kepler*-class photometry is investigated in this Chapter. Moreover, additional simulations of the exomoon light curve with intrinsic stellar noise were simulated to find out the effects of intrinsic stellar noise to the detectability of exomoons (Awiphan and Kerins 2013).

4.1 Exomoon Background and Detection Methods

Many of the planets of the Solar System host satellites. As the number of detected exoplanets continues to grow, the potential for detecting satellites orbiting them has become of increasing interest. The presence of exomoons may improve the probability of the existence of life on their host planets and the moons themselves also have potential to host life (Laskar et al. 1993). Moreover, the detection of exomoons would improve our understanding of planetary formation and evolution (Williams

© Springer International Publishing AG, part of Springer Nature 2018
S. Awiphan, *Exomoons to Galactic Structure*, Springer Theses,
https://doi.org/10.1007/978-3-319-90957-8_4

et al. 1997). In order to detect exomoons, several methods have been developed over the past decade.

Transit Method

The exomoon transit detection method is a detection method which detects the dip in the stellar light curve due the moon passing in front of the host star (Simon et al. 2007). However, it is very challenging to confirm the exomoon signal from the observational data, because the dip may occur due to some form of photometric noise or star spots (Sartoretti and Schneider 1999). The auxiliary transit, where the moon with a wide separation from the planet transits the star and creates the same transit shape as the planet, can occur where the moon locates ∼93% of the planet's Hill radius away from the planet (Domingos et al. 2006). The mutual events, where the moon and the planet separate at the start and end of the transit, but the moon eclipses the planet during the transit, can be detected when the exomoon is in a close-in orbit (5% of the Hill radius) (Heller et al. 2014). In 2018, Teachey et al. (2018) discovered an exomoon candidate, Kepler-1625b I, from three transits of Kepler-1625 b in the *Kepler* data. From the data, Kepler-1625b I is a Neptune-sized exomoon orbiting Kepler-1625b, Jupiter-sized exoplanet, with a separation of about 20 times the planetary radius.

Transit Timing Variation

A moon orbiting an exoplanet can cause a timing effect in the motion of the planet. This idea was proposed by Sartoretti and Schneider (1999). For a single moon system, the host planet and its exomoon orbit a common barycentre, which orbits the host star on a Keplerian orbit. Therefore, the planet itself does not orbit the host on a Keplerian orbit. The transits happen earlier or later depending on the phase of the exomoon. To date, many TTV theories which are extensions of the transit method have been proposed to detect exomoons, such as TTV (Sartoretti and Schneider 1999), the photocentric transit timing variation (PTV), in which the average of the light curve is measured (Szabó et al. 2006; Simon et al. 2007), and the transit duration variation (TDV) (Kipping 2009b) (See Sect. 4.3). Using TTV and TDV methods, exomoons should be detected with *Kepler* according to the simulation result of Kipping et al. (2009).

Microlensing

Han and Han (2002) proposed that exomoons could be detected using the microlensing technique. In 2014, Bennett et al. (2014) announced the first free-floating exomoon candidate, MOA-2011-BLG-262L b. From the best-fit solution of the event, a large relative proper motion was found which suggests a nearby lens system or a source star with very high proper motion. In the case of the nearby lens system, the best-fit solution shows that the system should contain a host planet with mass of ∼4 Jupiter masses and a sub-Earth mass moon. However, there is another solution with low relative proper motion which suggests a distant stellar host lens as well. As microlensing events cannot repeat, it is not possible to confirm an exomoon detection for this event. However, from this study, the result shows that the microlensing technique has the sensitivity required to detect exomoons. In the future, microlensing observations with the parallax technique can be used to measure the lens mass and confirm exomoon detections.

Direct Imaging

Direct imaging of exoplanets is extremely challenging, due to the difference in brightness between the host star and the planet. Detecting exomoons using the direct imaging technique is more difficult, because of a tiny angular separation between the planet and the moon. Normally, the brightness of exomoons is fainter than their host planets, due to their smaller sizes. For the Earth-moon system at a distance of 10 parsec, the separation is 0.5 milli-arcseconds, which is very small compared to the current best interferometric precision ($\sim$25 milliarcseconds) (Baines et al. 2007).

However, the exomoons orbiting Jovian-sized exoplanets in the habitable zones of main-sequence stars can be brighter than their host planets in the near-infrared (1–4 µm) in the cases where the planet atmospheres contain methane, water and water vapor that absorb light in that wavelength (Williams and Knacke 2004). Therefore, habitable zone Earth-sized moons may be detected by using that spectral contrast.

Pulsar Timing

The first exoplanet was discovered through the pulsar timing method. Lewis et al. (2008) proposed that this technique should be able to detect a stable exomoon orbiting a pulsar planet. They applied this method to the case of PSR B1620-26b and found that exomoons with mass larger than 5% of the pulsar planet and a planet-moon separation of 2% of star-planet separation could be detected.

Rossiter–McLaughlin Effect

The Rossiter–McLaughlin effect is a spectroscopic phenomenon observed when an exoplanet transits the host star. During a transit, part of the rotating host star surface is blocked, which caused a variation in Doppler shift (Rossiter 1924; McLaughlin 1924). Exomoon detection via the Rossiter–McLaughlin effect is proposed by Sumi (2010), Zhuang et al. (2012). The radius of the exomoon can be calculated from the half-amplitude of the Rossiter–McLaughlin effect, A_{RM}, and rotation velocity of the star, v_{rot},

$$A_{\mathrm{RM}} \propto \left(\frac{R_M}{R_*} \right)^2 v_{\mathrm{rot}} \sin i \, , \tag{4.1}$$

where R_M and R_* are the radius of the exomoon and the host star, respectively (Sumi 2010). Therefore, combining Rossiter–McLaughlin observation with transit data, the mass, radius and density of the exomoon can be determined.

Orbital Sampling Effect

The orbital sampling effect is the method where the photometric transit light curves of exoplanets with exomons are phase-folded (Heller 2014). The exomoons normally locate at large separation from their host planets. The phase-folded light curves can be used to detect the exomoon transits by accumulating all exomoon transits.

Transit of Exomoon Plasma Tori

The plasma tori produced by volcanically active exomoons around giant planets can be large enough to produce detectable transit absorption (Ben-Jaffel and Ballester

2014). Small exomoons can be detected, if they have enough volcanic activity to produce a spatially extended plasma nebula which may show in the transit at UV wavelengths.

Radio Emission

The exomoon motion can produce currents along the electromagnetic field lines which connect them to the Jovian host planet's polar regions. The currents generate and modulate radio emission along their paths via the electron-cyclotron maser instability which can be detected (Noyola et al. 2014).

4.2   Generating Light Curves of an Exoplanet with Exomoon

In this section, the coordinate system and fundamental parameters which form the basis of exomoon light curve calculations are defined. Transiting exoplanets are detected from the variation in the light curve of the host stars. Information from transit light curves helps astronomers to obtain direct estimates of the exoplanet radius which reveal the physical nature of the exoplanet. A very high precision light curve can be used to detect multi-planetary systems, star spots or exomoons (Charbonneau et al. 2007; Teachey et al. 2018).

4.2.1   Planet-Moon Orbit

The first step to generating a transiting exomoon light curve is to define the planet and moon positions with respect to the host star. However, the star, planet and moon are a three-body problem which has no exact analytical solution. Therefore, in following work, the planet's orbit and moon's orbit are created using a Keplerian orbit of star-planet and planet-moon, separately, and then perturbing the planet's position with planet-moon interaction.

Firstly, a planetary orbit with a centre at the origin of an (x_1, y_1, z_1) co-ordinate system is considered. The position of the planet can be written as,

$$x_1 = a_p e_p + r_p \cos f_{tp} \,,$$
$$y_1 = r_p \sin f_{tp} \,,$$
$$z_1 = 0 \,, \tag{4.2}$$

where a_p is planet's semi-major axis, e_p is planet's eccentricity, r_p is the distance of planet from the origin,

$$r_p = \frac{a_p(1 - e_p^2)}{1 + e_p} \cos(f_{tp}) \,, \tag{4.3}$$

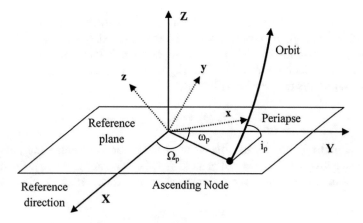

Fig. 4.1 Orbital elements of a transiting planet-star system

and true anomaly, f_{tp}, is defined by,

$$\tan\left(\frac{f_{tp}}{2}\right) = \sqrt{\frac{1+e_p}{1-e_p}} \cdot \tan\left(\frac{f_{mp}}{2}\right). \tag{4.4}$$

f_{mp} is the mean anomaly which may be written as a function of the time elapsed since the planet's periapsis, $t - t_p$,

$$f_{mp} - e_p \cos f_{mp} = \frac{2\pi}{P_p}(t - t_p). \tag{4.5}$$

In the second step, the coordinates (x_1, y_1, z_1) are transformed to star planet barycentre coordinates (x_2, y_2, z_2) lying at $(a_p e_p, 0, 0)$ in the (x_1, y_1, z_1) co-ordinate system:

$$
\begin{aligned}
x_2 &= r_p \cos(f_{tp}), \\
y_2 &= r_p \sin(f_{tp}), \\
z_2 &= 0.
\end{aligned} \tag{4.6}
$$

From Fig. 4.1, the orbital arrangement of the planet is rotated by the Euler angles: longitude of ascending node (Ω_p), inclination (i_p) and argument of periapsis (ω_p), with respect to the plane of reference (the reference direction lies at $x = +\infty$) which can be written as,

$$
\begin{aligned}
x_3 &= r_p \left[\cos(\Omega_p)\cos(\omega_p + f_{tp}) + \sin(i_p)\sin(\Omega_p)\sin(\omega_p + f_{tp})\right], \\
y_3 &= r_p \left[\sin(\Omega_p)\cos(\omega_p + f_{tp}) + \sin(i_p)\cos(\Omega_p)\sin(\omega_p + f_{tp})\right], \\
z_3 &= r_p \left[\cos(i_p)\sin(\omega_p + f_{tp})\right].
\end{aligned} \tag{4.7}
$$

If the planet has a moon, the moon's orbit can be defined in a similar manner to the planet's orbit. From here onwards, unless otherwise stated, quantities with subscript p denote planet parameters and those with subscript m denote moon parameters. The presence of the moon perturbs the planetary orbit. The planet's reflex motion from the moon's perturbation can be written as,

$$\mathbf{r_{bp}} M_p = \mathbf{r_{bm}} M_m , \tag{4.8}$$

where $\mathbf{r_{bp}}$ and $\mathbf{r_{bm}}$ are distance between planet-moon barycentre and planet and moon, respectively (where hereafter bold typeface denotes a vector). Finally, the position of the planet and moon with respect to star-planet reference plane, $\mathbf{r_{sp}}$ and $\mathbf{r_{sm}}$, are

$$\mathbf{r_{sp}} = \mathbf{r_p} + \mathbf{r_{bp}} , \tag{4.9}$$

$$\mathbf{r_{sm}} = \mathbf{r_p} + \mathbf{r_{bm}} . \tag{4.10}$$

For an observer at $z = +\infty$, Ω_p has no effect on the transit since the light curve is defined by the separation only. The planet position can be defined as,

$$\begin{aligned}
x_p = \; & r_p \cos(\omega_p + f_{tp}) \\
& -r_{bp} \cos(\omega_m + f_{tm}) \cos(\omega_p + \Omega_m) \\
& +r_{bp} \sin(i_m) \sin(\omega_m + f_{tm}) \sin(\omega_p + \Omega_m) ,
\end{aligned} \tag{4.11}$$

$$\begin{aligned}
y_p = \; & r_p \cos(i_p) \sin(\omega_p + f_{tp}) \\
& +r_{bp} \sin(i_p) \cos(i_m) \sin(\omega_m + f_{tm}) \\
& -r_{bp} \cos(i_p) \sin(i_m) \sin(\omega_m + f_{tm}) \cos(\omega_p + \Omega_m) \\
& -r_{bp} \cos(i_p) \cos(\omega_m + f_{tm}) \sin(\omega_p + \Omega_m) ,
\end{aligned} \tag{4.12}$$

$$\begin{aligned}
z_p = \; & r_p \sin(i_p) \sin(\omega_p + f_{tp}) \\
& -r_{bp} \cos(i_p) \cos(i_m) \sin(\omega_m + f_{tm}) \\
& -r_{bp} \sin(i_p) \sin(i_m) \sin(\omega_m + f_{tm}) \cos(\omega_p + \Omega_m) \\
& -r_{bp} \sin(i_p) \cos(\omega_m + f_{tm}) \sin(\omega_p + \Omega_m) ,
\end{aligned} \tag{4.13}$$

and moon position can be written as,

$$\begin{aligned}
x_m = \; & r_p \cos(\omega_p + f_{tp}) \\
& +r_{bm} \cos(\omega_m + f_{tm}) \cos(\omega_p + \Omega_m) \\
& -r_{bm} \sin(i_m) \sin(\omega_m + f_{tm}) \sin(\omega_p + \Omega_m) ,
\end{aligned} \tag{4.14}$$

$$y_m = r_p \cos(i_p) \sin(\omega_p + f_{tp})$$
$$-r_{bm} \sin(i_p) \cos(i_m) \sin(\omega_m + f_{tm})$$
$$+r_{bm} \cos(i_p) \sin(i_m) \sin(\omega_m + f_{tm}) \cos(\omega_p + \Omega_m)$$
$$+r_{bm} \cos(i_p) \cos(\omega_m + f_{tm}) \sin(\omega_p + \Omega_m) \,, \tag{4.15}$$

$$z_m = r_p \sin(i_p) \sin(\omega_p + f_{tp})$$
$$+r_{bm} \cos(i_p) \cos(i_m) \sin(\omega_m + f_{tm})$$
$$+r_{bm} \sin(i_p) \sin(i_m) \sin(\omega_m + f_{tm}) \cos(\omega_p + \Omega_m)$$
$$+r_{bm} \sin(i_p) \cos(\omega_m + f_{tm}) \sin(\omega_p + \Omega_m) \,. \tag{4.16}$$

The star-moon sky-projected distance (S_{sm}), the star-planet sky-projected distance (S_{sp}) and the separation between the planet and the moon (S_{pm}) can be written as,

$$S_{sm}^2 = \frac{x_m^2 + y_m^2}{R_*^2} \,, \tag{4.17}$$

$$S_{sp}^2 = \frac{x_p^2 + y_p^2}{R_*^2} \,, \tag{4.18}$$

$$S_{pm}^2 = \frac{(x_p - x_m)^2 + (y_p - y_m)^2}{R_*^2} \,, \tag{4.19}$$

where R_* is star radius.

4.2.2 Planetary Transit Light Curve

In order to generate light curves of transiting systems, the limb darkening effect of the host star which causes the star's surface brightness peak at the centre is considered. The nonlinear limb-darkening in a transit light curve which is first presented by Mandel and Agol (2002). The function of the star's intensity, I, with nonlinear limb-darkening is defined by (Fig. 4.2),

$$I(r) = 1 + \sum_{n=1}^{4} c_n (1 - \mu^{n/2}) \,, \tag{4.20}$$

where c_n are coefficients and μ is the normalised radial coordinate of the star,

$$\mu = \cos\theta = (1 - r^2)^{1/2}, 0 \le r \le 1 \,, \tag{4.21}$$

Fig. 4.2 Geometry of limb darkening

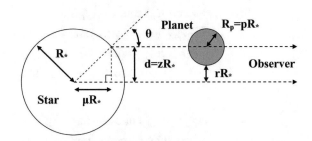

where θ is the angle between a line normal to the stellar surface and the line of sight of the observer and r is radial distance from the centre of the star. In this dissertation, the quadratic limb darkening, $I(r) = 1 - \gamma_1(1 - \mu) - \gamma_2(1 - \mu)^2$, where $\gamma_1 + \gamma_2 < 1$, is used. From Eq. 4.20, the constant c_n in the quadratic limb darkening model can be written as,

$$c_1 = 0$$
$$c_2 = \gamma_1 + 2\gamma_2$$
$$c_3 = 0$$
$$c_4 = -\gamma_2 , \tag{4.22}$$

and

$$c_0 \equiv 1 - c_1 - c_2 - c_3 - c_4 . \tag{4.23}$$

In what follows, R_* is the star radius, R_p is the planet radius, d is the center-to-center distance between the star and the planet, $z = d/R_*$ is the normalized separation, and $p = R_p/R_*$ is the size ratio. The light curve with quadratic limb darkening of the system can be written as,

$$F = 1 - \frac{1}{4\Omega} \left\{ (1 - c_2)\lambda^e + c_2 \left[\lambda^d + \frac{2}{3}\Theta(p - z) \right] - c_4\eta^d \right\} . \tag{4.24}$$

where $\Omega = \sum_{n=0}^{4} c_n (n + 4)^{-1}$ and $\Theta(p - z)$ is heaviside step function,

$$\Theta(p - z) = \begin{cases} 0, & p - z < 0 , \\ \frac{1}{2}, & p - z = 0 , \\ 1, & p - z > 1 , \end{cases} \tag{4.25}$$

and λ^e is defined by

$$\lambda^e(p, z) = \begin{cases} 0, & 1 + p < z , \\ \frac{1}{\pi}(p^2\kappa_0 + \kappa_1 - \kappa_2), & |1 - p| < z \leq 1 + p , \\ p^2, & z \leq 1 - p , \\ 1, & z \leq p - 1 , \end{cases} \tag{4.26}$$

$$\kappa_0 = \cos^{-1} \frac{-1 + p^2 + z^2}{2pz}, \tag{4.27}$$

$$\kappa_1 = \cos^{-1} \frac{1 - p^2 + z^2}{2z}, \tag{4.28}$$

$$\kappa_2 = \sqrt{\frac{4z^2 - (1 + z^2 - p^2)^2}{4}}. \tag{4.29}$$

Finally, the value of λ^d and η^d are defined in Table 4.1 and Eq. 4.30 (Mandel and Agol 2002).

$$
\begin{aligned}
\lambda_1 &= \frac{1}{9\pi\sqrt{pz}} \left[(1-b)(2b+a-3) - 3q(b-2) \right] K(k) \\
&\quad + \frac{4pz}{9\pi\sqrt{pz}} \left[(z^2 + 7p^2 - 4) \right] E(k) - \frac{3}{9\pi\sqrt{pz}} \left[\frac{q}{a} \Pi\left(\frac{a-1}{a}, k \right) \right], \\
\lambda_2 &= \frac{2}{9\pi\sqrt{1-a}} \left[q^2 K(k^{-1}) + (1-a)(z^2 + 7p^2 - 4) E(k^{-1}) \right] \\
&\quad + \frac{2}{9\pi\sqrt{1-a}} \left[1 - 5z^2 + p^2 \right] - \frac{2}{3\pi\sqrt{1-a}} \left[\frac{q}{a} \Pi\left(\frac{a-b}{a}, k^{-1} \right) \right], \\
\lambda_3 &= \frac{1}{3} + \frac{16p}{9\pi} \left[(2p^2 - 1) E\left(\frac{1}{2k} \right) \right] - \frac{(1-4p^2)(3-8p^2)}{9\pi p} K\left(\frac{1}{2k} \right), \\
\lambda_4 &= \frac{1}{3} + \frac{2}{9\pi} \left[4(2p^2 - 1)E(2k) + (1-4p^2)K(2k) \right], \\
\lambda_5 &= \frac{2}{3\pi} \left[\cos^{-1}(1 - 2p) \right] - \frac{4}{9\pi} \left[3 + 2p - 8p^2 \right], \\
\lambda_6 &= -\frac{2}{3} \left[1 - p^2 \right]^{\frac{3}{2}}, \\
\eta_1 &= \frac{1}{2\pi} \left[\kappa_1 + 2\eta_2\kappa_0 - \frac{1}{4}(1 + 5p^2 + z^2)\sqrt{(1-a)(b-1)} \right], \\
\eta_2 &= \frac{p^2}{2} \left[p^2 + 2z^2 \right],
\end{aligned}
\tag{4.30}
$$

where $E(k)$, $K(k)$ and $\Pi(n, k)$ is the complete elliptic integral of the first kind, second kind and third kind, respectively.

$$E(k) = \int_0^{\frac{\pi}{2}} \frac{1}{\sqrt{1 - k^2 \sin^2(x)}} dx \tag{4.31}$$

$$K(k) = \int_0^{\frac{\pi}{2}} \sqrt{1 - k^2 \sin^2(x)} dx \tag{4.32}$$

Table 4.1 Limb-darkening occultation

Case	p	z	$\lambda^d(z)$	$\eta^d(z)$
1	$(0, \infty)$	$[1 + p, \infty)$	0	0
	0	$[0, \infty)$	0	0
2	$(0, \infty)$	$(\frac{1}{2} + \lvert p - \frac{1}{2} \rvert, 1 + p)$	λ_1	η_1
3	$(0, \frac{1}{2})$	$(p, 1 - p)$	λ_2	η_2
4	$(0, \frac{1}{2})$	$1 - p$	λ_5	η_2
5	$(0, \frac{1}{2})$	p	λ_4	η_2
6	$\frac{1}{2}$	$\frac{1}{2}$	$\frac{1}{3} - \frac{4}{9}\pi$	$\frac{3}{32}$
7	$(\frac{1}{2}, \infty)$	p	λ_3	η_1
8	$(\frac{1}{2}, \infty)$	$[\lvert 1 - p \rvert, p)$	λ_1	η_1
9	$(0, 1)$	$(0, \frac{1}{2} - \lvert p - \frac{1}{2} \rvert)$	λ_2	η_2
10	$(0, 1)$	0	λ_6	η_2
11	$(1, \infty)$	$[0, p - 1)$	1	1

$$\Pi(n, k) = \int_0^{\frac{\pi}{2}} \frac{1}{(1 - n \sin^2(x))\sqrt{1 - k^2 \sin^2(x)}} dx \qquad (4.33)$$

4.2.3 Planet-Moon Transit Light Curve

In following work, LUNA algorithm was used to generating light curves of an exoplanet with exomoon (Kipping 2011a). For the moon which transit the star with the area, $A_{m,t}$, its actively transit component is assumed to equal to the actively transit component of the planet with equal area. Therefore, the flux due to the planet-moon projection, $F_{transit}$, can be written as,

$$F_{transit} = F_{planet} - A_{m,t} \frac{F_{area}}{F_{total}} , \qquad (4.34)$$

where F_{planet} is the flux of planetary transit and F_{total} is the total stellar flux,

$$
\begin{aligned}
F_{total} &= \int_0^1 2r I(r) dr \\
&= 1 - \sum_{n=1}^4 \frac{n c_n}{n + 4} \\
&= 1 - \frac{1}{5}c_1 - \frac{1}{3}c_2 - \frac{3}{7}c_3 - \frac{1}{2}c_4 , \qquad (4.35)
\end{aligned}
$$

where $I(r)$ is the intensity at the center-to-center distance between the star and the planet, r. The value of area flux, F_{area}, and ratio of area, A_m, depend on the moon-star separation (Table 4.2).

Case I

If the moon is outside the star. The area flux and ratio of area are set to be zero.

Case II

If the moon is in the ingress or egress portion, $1 - s < S_{sm} < 1 + s$, then,

$$
\begin{aligned}
F_{area,II} &= \int_{S_{sm}-s}^{1} 2r I(r) dr \\
&= (a_m - 1)(c_1 + c_2 + c_3 + c_4 - 1) \\
&\quad + \frac{4}{5}c_1(1 - a_m)^{\frac{5}{4}} + \frac{2}{3}c_2(1 - a_m)^{\frac{3}{2}} \\
&\quad + \frac{4}{7}c_3(1 - a_m)^{\frac{7}{4}} + \frac{1}{2}c_4(1 - a_m)^2 ,
\end{aligned}
\tag{4.36}
$$

where $a_m = (S_{sm} - s)^2$ and $s = R_m/R_*$ is the moon size ratio.

Case III

If the moon is inside the star but does not cover the centre of the star, $s < S_{sm} < 1 - s$, then,

Table 4.2 Moon active transit component

Case	Condition	Area flux (F_{area})	Ratio of Area (A_m)
I	$1 + s < S_{sm} < \infty$	0	0
II	$1 - s < S_{sm} < 1 + s$	Equation 4.36	$A_{m,t}/(\pi(1 - a_m))$
III	$s < S_{sm} < 1 - s$	Equation 4.37	$A_{m,t}/(\pi(b_m - a_m))$
IV	$0 < S_{sm} < s$	Equation 4.38	$A_{m,t}/(\pi b_m)$

$$F_{area,III} = \int_{S_{sm}-s}^{S_{sm}+s} 2r\,I(r)\,dr$$

$$= (a_m - b_m)(c_1 + c_2 + c_3 + c_4 - 1)$$

$$+ \frac{4}{5}c_1(1 - a_m)^{\frac{5}{4}} + \frac{2}{3}c_2(1 - a_m)^{\frac{3}{2}}$$

$$+ \frac{4}{7}c_3(1 - a_m)^{\frac{7}{4}} + \frac{1}{2}c_4(1 - a_m)^2$$

$$- \frac{4}{5}c_1(1 - b_m)^{\frac{5}{4}} - \frac{2}{3}c_2(1 - b_m)^{\frac{3}{2}}$$

$$- \frac{4}{7}c_3(1 - b_m)^{\frac{7}{4}} - \frac{1}{2}c_4(1 - b_m)^2 , \qquad (4.37)$$

where $b_m = (S_{sm} + s)^2$.

Case IV

If the moon is inside the star and cover the centre of the star, $0 < S_{sm} < s$, then,

$$F_{area,IV} = \int_0^{S_{sm}+s} 2r\,I(r)\,dr$$

$$= -b_m(c_1 + c_2 + c_3 + c_4 - 1)$$

$$+ \frac{4}{5}c_1 + \frac{2}{3}c_2 + \frac{4}{7}c_3 + \frac{1}{2}c_4$$

$$- \frac{4}{5}c_1(1 - b_m)^{\frac{5}{4}} - \frac{2}{3}c_2(1 - b_m)^{\frac{3}{2}}$$

$$- \frac{4}{7}c_3(1 - b_m)^{\frac{7}{4}} - \frac{1}{2}c_4(1 - b_m)^2 . \qquad (4.38)$$

The Actively Transiting Area of Moon

The actively transiting area of moon can be described by 3 parameters; S_{sp}, S_{sm} and S_{pm}. There are 27 principal cases including some unphysical cases which listed in Table 4.3 and Fig. 4.3 (Kipping 2011a).

For some cases, the transiting areas are described by the area of intersection between any two circles, α. The area of transit caused by an object of radius r transiting an object of radius R, with separation S, is

$$\alpha(R, r, S) = r^2\kappa_0(R, r, S) + R^2\kappa_1(R, r, S) - \kappa_2(R, r, S) , \qquad (4.39)$$

$$\kappa_0(R, r, S) = \cos^{-1}\frac{S^2 + r^2 - R^2}{2Sr} , \qquad (4.40)$$

$$\kappa_1(R, r, S) = \cos^{-1}\frac{S^2 - r^2 + R^2}{2SR} , \qquad (4.41)$$

$$\kappa_2(R, r, S) = \sqrt{\frac{4S^2R^2 - (R^2 + S^2 - r^2)^2}{4}} . \qquad (4.42)$$

Table 4.3 The actively transiting area of planet and moon

Case	S_{sp}	S_{sm}	S_{pm}	Physical	$A_{p,t}$	$A_{m,t}$
1	$S_{sp} \geq 1+p$	$S_{sm} \geq 1+s$	$S_{pm} \geq p+s$	✓	0	0
2	$S_{sp} \geq 1+p$	$S_{sm} \geq 1+s$	$p-s < S_{pm} < p+s$	✓	0	0
3	$S_{sp} \geq 1+p$	$S_{sm} \geq 1+s$	$S_{pm} \leq p-s$	✓	0	0
4	$S_{sp} \geq 1+p$	$1-s < S_{sm} < 1+s$	$S_{pm} \geq p+s$	✓	0	α_{sm}
5	$S_{sp} \geq 1+p$	$1-s < S_{sm} < 1+s$	$p-s < S_{pm} < p+s$	✓	0	α_{sm}
6	$S_{sp} \geq 1+p$	$1-s < S_{sm} < 1+s$	$S_{pm} \leq p-s$	✗	–	–
7	$S_{sp} \geq 1+p$	$S_{sm} \leq 1-s$	$S_{pm} \geq p+s$	✓	0	πs^2
8	$S_{sp} \geq 1+p$	$S_{sm} \leq 1-s$	$p-s < S_{pm} < p+s$	✗	–	–
9	$S_{sp} \geq 1+p$	$S_{sm} \leq 1-s$	$S_{pm} \leq p-s$	✗	–	–
10	$1-p < S_{sp} < 1+p$	$S_{sm} \geq 1+s$	$S_{pm} \geq p+s$	✓	α_{sp}	0
11	$1-p < S_{sp} < 1+p$	$S_{sm} \geq 1+s$	$p-s < S_{pm} < p+s$	✓	α_{sp}	0
12	$1-p < S_{sp} < 1+p$	$S_{sm} \geq 1+s$	$S_{pm} \leq p-s$	✓	α_{sp}	0
13	$1-p < S_{sp} < 1+p$	$1-s < S_{sm} < 1+s$	$S_{pm} \geq p+s$	✓	α_{sp}	α_{sm}
14	$1-p < S_{sp} < 1+p$	$1-s < S_{sm} < 1+s$	$p-s < S_{pm} < p+s$	✓	α_{sp}	§Fewell
15	$1-p < S_{sp} < 1+p$	$1-s < S_{sm} < 1+s$	$S_{pm} \leq p-s$	✓	α_{sp}	0
16	$1-p < S_{sp} < 1+p$	$S_{sm} \leq 1-s$	$S_{pm} \geq p+s$	✓	α_{sp}	πs^2
17	$1-p < S_{sp} < 1+p$	$S_{sm} \leq 1-s$	$p-s < S_{pm} < p+s$	✓	α_{sp}	$\pi s^2 - \alpha_{pm}$
18	$1-p < S_{sp} < 1+p$	$S_{sm} \leq 1-s$	$S_{pm} \leq p-s$	✓	α_{sp}	0
19	$S_{sp} \leq 1-p$	$S_{sm} \geq 1+s$	$S_{pm} \geq p+s$	✓	πp^2	0
20	$S_{sp} \leq 1-p$	$S_{sm} \geq 1+s$	$p-s < S_{pm} < p+s$	✓	–	0
21	$S_{sp} \leq 1-p$	$S_{sm} \geq 1+s$	$S_{pm} \leq p-s$	✓	–	–
22	$S_{sp} \leq 1-p$	$1-s < S_{sm} < 1+s$	$S_{pm} \geq p+s$	✓	πp^2	α_{sm}
23	$S_{sp} \leq 1-p$	$1-s < S_{sm} < 1+s$	$p-s < S_{pm} < p+s$	✓	πp^2	$\alpha_{sm} - \alpha_{pm}$
24	$S_{sp} \leq 1-p$	$1-s < S_{sm} < 1+s$	$S_{pm} \leq p-s$	✓	–	–
25	$S_{sp} \leq 1-p$	$S_{sm} \leq 1-s$	$S_{pm} \geq p+s$	✓	α_{sp}	πs^2
26	$S_{sp} \leq 1-p$	$S_{sm} \leq 1-s$	$p-s < S_{pm} < p+s$	✓	α_{sp}	$\pi s^2 - \alpha_{pm}$
27	$S_{sp} \leq 1-p$	$S_{sm} \leq 1-s$	$S_{pm} \leq p-s$	✓	α_{sp}	0

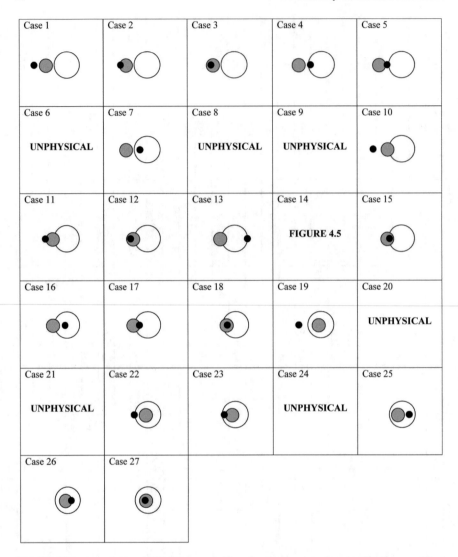

Fig. 4.3 Diagrams of cases 1–27 show star (white sphere), planet (gray sphere) and moon (black sphere)

Case 14 (Fewell Case)

Case 14 is the most complicated case to consider, because the planet's shadow and moon's shadow do not completely eclipse the star and there is possibility that they overlap each other. Therefore, the transiting area is described by area of intersection of three circles. Fewell (2006) presented the solution showing that the intersection points of three circles can be written as,

Star-Planet Intersection

$$x_{sp} = \frac{1 - p^2 + S_{sp}^2}{2S_{sp}} \, . \tag{4.43}$$

$$y_{sp} = \frac{1}{2S_{sp}} \sqrt{2S_{sp}^2(1 + p^2) - (1 - p^2)^2 - S_{sp}^4} \, . \tag{4.44}$$

Star-Moon Intersection

$$x_{sm} = x'_{sm} \cos\theta' - y'_{sm} \sin\theta' \, . \tag{4.45}$$

$$y_{sm} = x'_{sm} \sin\theta' + y'_{sm} \cos\theta' \, . \tag{4.46}$$

where

$$x'_{sm} = \frac{1 - s^2 + S_{sm}^2}{2S_{sm}} \, . \tag{4.47}$$

$$y'_{sm} = \frac{-1}{2S_{sm}} \sqrt{2S_{sm}^2(1 + s^2) - (1 - s^2)^2 - S_{sm}^4} \, . \tag{4.48}$$

$$\cos\theta' = \frac{S_{sp}^2 + S_{sm}^2 - S_{pm}^2}{2S_{sp}S_{sm}} \, . \tag{4.49}$$

$$\sin\theta' = \sqrt{1 - \cos^2\theta'} \, . \tag{4.50}$$

Planet-Moon Intersection

$$x_{pm} = x''_{pm} \cos\theta'' - y''_{pm} \sin\theta'' + S_{sp} \, . \tag{4.51}$$

$$y_{pm} = x''_{pm} \sin\theta'' + y''_{pm} \cos\theta'' \, . \tag{4.52}$$

where

$$x''_{pm} = \frac{p^2 - s^2 + S_{pm}^2}{2S_{pm}} \, . \tag{4.53}$$

$$y''_{pm} = \frac{1}{2S_{pm}} \sqrt{2S_{pm}^2(p^2 + s^2) - (p^2 - s^2)^2 - S_{pm}^4} \, . \tag{4.54}$$

$$\cos\theta'' = -\frac{S_{sp}^2 + S_{pm}^2 - S_{sm}^2}{2S_{sp}S_{pm}} \, . \tag{4.55}$$

$$\sin\theta'' = \sqrt{1 - \cos^2\theta''} \, . \tag{4.56}$$

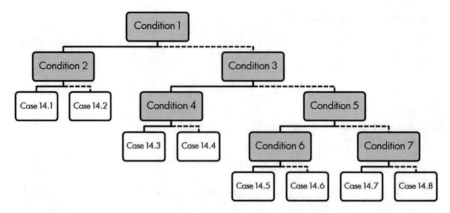

Fig. 4.4 Decision flow chart of Case 14. Gy boxes are the decision conditions and white boxes are subcases of Case 14. The solid lines indicate a true statement and the dashed lines indicate a false statement

In order to calculate the transit area of the moon, there are seven conditions that the simulation may take into account. The list of conditions is shown in Eq. 4.57.

Condition 1 $(x_{sp} - S_{sm} \cos \theta')^2 + (y_{sp} + S_{sm} \sin \theta')^2 < s^2$

Condition 2 $\qquad\qquad S_{sp} > 1$

Condition 3 $(x_{sp} - S_{sm} \cos \theta')^2 + (y_{sp} - S_{sm} \sin \theta')^2 < s^2$

Condition 4 $S_{sm} \sin \theta' > y_{sm} + \frac{y_{pm} - y_{sm}}{x_{pm} - x_{sm}} (S_{sm} \cos \theta' - x_{sm})$

Condition 5 $\qquad\qquad (x_{sm} - S_{sp})^2 + y_{sm}^2 < p^2$

Condition 6 $\qquad\qquad (S_{sm} - s) < (S_{pm} - p)$

Condition 7 $\qquad\qquad x_{pm}^2 + y_{pm}^2 < 1$

The decision flow chart for calculating moon's transiting area of Case 14 is shown in Fig. 4.4. In Table 4.4 and Fig. 4.5, the transiting areas of the moon in each subcase of Case 14 are listed.

$$
\begin{aligned}
\alpha_{14.3} = {} & \frac{1}{4}\sqrt{(c_1 + c_2 + c_3)(-c_1 + c_2 + c_3)(c_1 - c_2 + c_3)(c_1 + c_2 - c_3)} \\
& + \sum_{k=1}^{3}\left(R_k^2 \arcsin \frac{c_k}{2R_k} \right) - \frac{c_1}{4}\sqrt{4R_1^2 - c_1^2} \\
& - \frac{c_2}{4}\sqrt{4R_2^2 - c_2^2} - \frac{c_3}{4}\sqrt{4R_3^2 - c_1^3},
\end{aligned}
\tag{4.57}
$$

and

Table 4.4 Moon active transit component of Case 14 (Kipping 2011a)

Case	Ratio of Area ($\mathbf{A_m}$)
14.1	$\alpha_{sm} - \alpha_{sp}$
14.2	$\pi p^2 - \alpha_{pm} - \alpha_{sp} + \alpha_{sm}$
14.3	$\alpha_{sm} - \alpha_{14.3}$
14.4	$\alpha_{sm} - \alpha_{14.4}$
14.5	$\pi s^2 - \alpha_{pm}$
14.6	0
14.7	α_{sm}
14.8	$\alpha_{sm} - \alpha_{pm}$

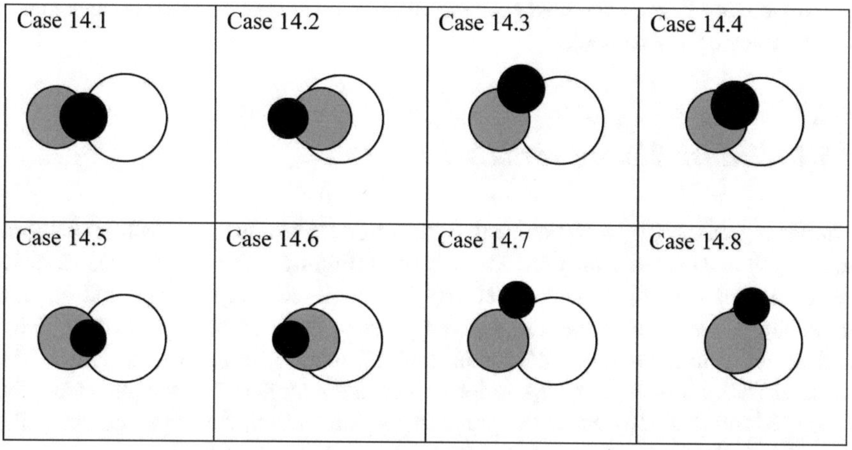

Fig. 4.5 Diagrams of cases 14 show star (white sphere), planet (gray sphere) and moon (black sphere)

$$\alpha_{14.4} = \frac{1}{4}\sqrt{(c_1 + c_2 + c_3)(-c_1 + c_2 + c_3)(c_1 - c_2 + c_3)(c_1 + c_2 - c_3)}$$

$$+ \sum_{k=1}^{3}\left(R_k^2 \arcsin\frac{c_k}{2R_k}\right) - \frac{c_1}{4}\sqrt{4R_1^2 - c_1^2}$$

$$- \frac{c_2}{4}\sqrt{4R_2^2 - c_2^2} + \frac{c_3}{4}\sqrt{4R_3^2 - c_1^3}, \tag{4.58}$$

where R_k is radius of object and c_k is chord lengths,

$$c_k^2 = (x_{ik} - x_{jk})^2 + (y_{ik} - y_{jk})^2 . \tag{4.59}$$

4.3 Transit Timing Variations and Transit Duration Variations

In order to detect exomoons, several methods have been developed as outlined in Sect. 4.1. One of the most efficient methods is the TTV method proposed by Sartoretti and Schneider (1999). They suggested that a moon orbiting an exoplanet can be detected by measuring the planet transit timing variation, due to the gravitational interaction between moon and planet. The moon's gravity causes the planet to orbit around planet-moon barycentre. Therefore, during the transit, the position relative to the barycentre of the planet changes. Moreover, the TDV signal can occur where the planet's velocity can be measured by changes in transit duration. Like the Doppler spectroscopic effect of the host star, the velocity of the planet also changes due to the presence of an exomoon.

4.3.1 Transit Timing Variations

The concept of the TTV technique is that the presence of a third body such as exomoon in the system causes a change in planetary orbit (Fig. 4.6). The time between transits varies because the transiting planet and the moon exchange energy and angular momentum. This gravitational interaction perturbs the orbit of the transiting planet and causes a short-period oscillation of the semi-major axes and eccentricities. The signal depends on the mass, separation and orbital parameters of the planet and the moon (Kipping 2011b). From the gravitational interaction, the displacement of the planet from the planet-moon barycentre is given by,

$$a_{pb} = \left(\frac{M_m}{M_p} \right) a_m , \qquad (4.60)$$

where a_{pb} is the semi-major axis of the planet around the planet-moon barycentre (Fig. 4.7). For edge-on circular orbits, the TTV signal (S_{TTV}) and root-mean-square (RMS) amplitude of the signal (δ_{TTV}) can be written as,

$$S_{TTV} = \left[\frac{a_m M_m P_p}{2\pi a_p M_p} \right] \sin(f_m) , \qquad (4.61)$$

and

$$\delta_{TTV} = \frac{a_m M_m P_p}{2\pi a_p M_p \sqrt{2}} , \qquad (4.62)$$

where M_m is the moon's mass and a_m is the semi-major axis of the moon around the planet-moon barycentre, P_p is the orbital period of the planet and f_m is the moon phase, $f_m = 0°$ when the moon is in opposition to the star. The peak-to-peak

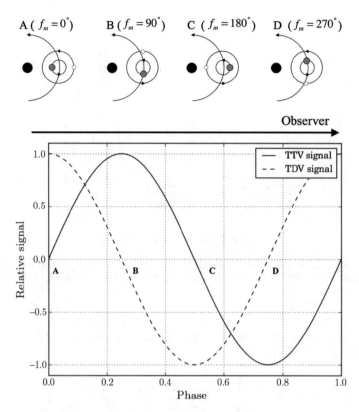

Fig. 4.6 Variation of TTV and TDV signals with planet (gray circle) and moon (white circle) positions

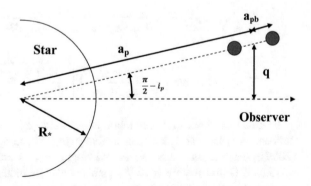

Fig. 4.7 The side-on view of the star-planet-moon system shows the distance perturbation of the planet caused by the moon (two grey circles)

amplitude of the TTV signal (Δt_{TTV}) can be written as,

$$\Delta t_{TTV} \sim 2a_m M_m M_p^{-1} \times P_p (2\pi a_p)^{-1} . \tag{4.63}$$

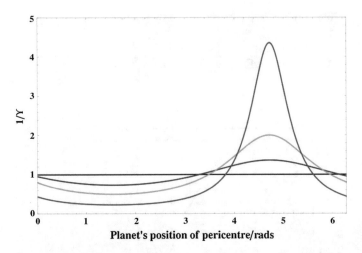

Fig. 4.8 The value of Υ^{-1} versus the planet's position of pericentre, ϖ_p. The black, blue, green and red lines represent planetary eccentricities (e_p) of 0, 0.3, 0.6 and 0.9 respectively (Kipping 2009a)

In the case of eccentric orbits, the waveforms are not sinusoidal. The RMS amplitude of the TTV signal is,

$$\delta_{TTV} = \frac{1}{\sqrt{2}} \frac{a_p^{1/2} a_m M_m (M_p + M_m)^{-1}}{\sqrt{G(M_* + M_p + M_m)}} \frac{\zeta_T(e_m, \varpi_m)}{\Upsilon(e_p, \varpi_p)} , \qquad (4.64)$$

where

$$\zeta_T = \frac{(1 - e_m^2)^{1/4}}{e_m} \sqrt{e_m^2 + \cos(2\varpi_m)(2(1 - e_m^2)^{3/2} - 2 + 3e_m^2)} , \qquad (4.65)$$

$$\Upsilon = \cos \left[\arctan \left(\frac{-e_p \cos \varpi_p}{1 + e_p \sin \varpi_p} \right) \right] \cdot \sqrt{\frac{2(1 + e_p \sin \varpi_p)}{1 - e_p^2} - 1} . \qquad (4.66)$$

In Eqs. 4.64–4.66, e_m is the moon eccentricity. ϖ_p and ϖ_m are the positions of pericentre of planet and moon, respectively (Fig. 4.8, Kipping 2009a). From Kipping (2009a), Earth-mass exomoons around Neptune-mass exoplanets could be detected by *Kepler*. Unfortunately, a TTV signal can also be induced by a multitude of phenomena, including general relativistic precession rate of periastra (Jordán and Bakos 2008), stellar proper motion (Rafikov 2009) and parallax (Scharf 2007) effects. Therefore, a TTV signal by itself cannot confirm the presence of an exomoon.

4.3.2 Transit Duration Variation

Kipping (2009a) showed that exomoons should induce not only the TTV effect but also the TDV effect on their host planets. TDV is the periodic change in the transit duration over many measurements caused by the apparent velocity of the planet which increases and decreases due to the planet-moon interaction. This has some similarity to the Doppler spectroscopy technique, but TDV observations involve tangential velocity variation rather than radial velocity.

For the systems with non-coplanar orbits, the TDV effect can be separated into two main constituents, a velocity (V) component and a transit impact parameter (TIP) component (Kipping 2009b). The V-component is caused by the variation in velocity of the planet due to the moon's gravity. The TIP-component is affected by the planet moving between high and low host-star impact parameters. Kipping (2011b) formulated that the total TDV signal is a linear combination between TDV-V signal and TDV-TIP signals (Kipping 2011b).

As TDV-V and TDV-TIP have 0 or π-phase difference, the RMS amplitude of the TDV signal is given by,

$$
\delta_{TDV} = \left[\underbrace{\frac{a_{pb} a_p \cos^2 i}{(R_* + R_P)^2 - a_p^2 \cos^2 i}}_{TIP-Component} \pm \underbrace{\frac{2\pi a_{pb}}{P_m} \frac{1}{v_{B\perp}}}_{V-Component} \right] \cdot \frac{\bar{\tau}}{\sqrt{2}} , \tag{4.67}
$$

where P_m is moon period, a_{pb} is the semi-major axis of the planet around the planet-moon barycentre, and $v_{B\perp}$ is the projected velocity of the planet-moon barycentre across the face of the star during transit. The positive sign refers to prograde moon orbits and the negative sign refers to retrograde orbits. For the planet's inclination of 90° and the moon's orbit which is coplanar with the planet-star orbit case, TDV-TIP vanishes and the TDV amplitude can be written as,

$$
\delta_{TDV} = \sqrt{\frac{a_p}{a_m}} \cdot \sqrt{\frac{M_m^2}{(M_p + M_m)(M_* + M_p + M_m)}} \cdot \frac{\bar{\tau}}{\sqrt{2}} \frac{\zeta_D(e_m, \varpi_m)}{\Upsilon(e_p, \varpi_p)} , \tag{4.68}
$$

$$
\delta_{TDV} \propto M_m a_m^{-1/2} , \tag{4.69}
$$

where

$$
\zeta_D = \sqrt{\frac{1 + e_m^2 - e_m^2 \cos(2\varpi_m)}{1 - e_m^2}} , \tag{4.70}
$$

with $\bar{\tau}$ is the duration of a transit ($\bar{\tau} \propto 1/v_{p\perp}$). Therefore, the TDV signal for an edge-on circular orbit is,

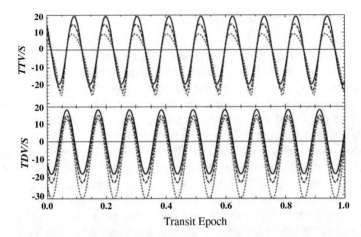

Transit Epoch

Fig. 4.9 The TTV and TDV signal simulated from 1 Earth-mass exomoon orbiting GJ436b. TTV leads TDV by a $\pi/2$ phase difference. The solid line represents $e_p = 0$, the dashed line $e_p = 0.3$ and the dotted line $e_p = 0.6$ (Kipping 2009a)

$$S_{TDV} = \bar{\tau} \left[\frac{a_m M_m P_p}{a_p M_p P_m} \right] \cos(f_m) , \qquad (4.71)$$

and the RMS amplitude of the TDV signal is,

$$\delta_{TDV} = \bar{\tau} \frac{a_m M_m P_p}{a_p M_p P_m \sqrt{2\pi}} . \qquad (4.72)$$

The TDV technique cannot detect habitable exomoons alone because the TDV signals are relatively weak compared with the TTV signals (Porter and Grundy 2011) and can also be induced by parallax effects (Scharf 2007). However, combining TDV and TTV signals can confirm the presence of an exomoon, because the signals have a $\frac{\pi}{2}$-phase difference that provides a unique exomoon signature (Fig. 4.9). Moreover, the orbital separation and mass of exomoons can also be obtained.

4.4 Habitable Exomoons

The search for and study of exomoons is interesting because the presence of exomoons may improve the probability of the existence of life on their host planets, and the exomoons themselves also have a potential to host life. In the case of the Moon, it has a stabilising effect on Earth's axis and causes the tides on the Earth. Without the Moon, life on the Earth may be restricted to less complex forms (Laskar et al. 1993). The presence of large exomoons may also improve the probability of the life on their host planets, as a large moon can lock the planet and moon together, preventing tidal locking to the star.

Moreover, a large moon could be habitable itself. Williams et al. (1997) suggested that the mass of a habitable moon is larger than $0.12\ M_\oplus$. Chyba (1997) and Williams et al. (1997) also proposed that exomoons orbiting inside the giant gas planets' Hill sphere (the region of orbital stability) could host life. Although the moons orbiting giant planets at 1 AU from a solar analogue would become tidally locked within a few billion years after they form, their orbital period on timescales of a few days to a few months could cause temperature fluctuations on them.

Habitable moons need to be large enough to retain water and an atmosphere, as Williams et al. (1997) suggested. Although moons formed from the planetary disk are unlikely to be greater than 0.01% the mass of the host planet, moons formed from captures or impacts (Triton and the Moon) can be greater than 0.01% the mass of the host planet (Canup and Ward 2006).

In order to find habitable exomoons, they should be located in the habitable zone of the system. The habitable zone of exomoons can be defined simply as the distance where planets receive the same energy as the Earth, r_{hab},

$$r_{hab} = \sqrt{\frac{L_*}{L_\odot}}\ \text{AU}\ , \tag{4.73}$$

where $L_\odot$ is the Sun's luminosity (Kipping et al. 2009). Furthermore, the moon can be lost from the planet due to three-body instability if the distance between the planet and moon is too large. To be retained by the planet, the moon must have an orbit that lies within the Hill sphere. The Hill sphere is a region that approximates the gravitational sphere of influence. The radius of the sphere, called the Hill radius (R_H), is found by solving the three-body problem and is equal to the distance of the $L1$ and $L2$ Lagrangian points, which lie along the line of centres of the two bodies.

$$R_H = a_p \left(\frac{M_p}{3M_*} \right)^{1/3}. \tag{4.74}$$

Barnes and O'Brien (2002) approximated the stability for prograde and retrograde moons as,

$$a_m \leq 0.36R_H\ , \tag{4.75}$$

and

$$a_m \leq 0.50R_H\ , \tag{4.76}$$

respectively. For exomoon orbital eccentricity, e_m, and planetary eccentricity, e_p, Domingos et al. (2006) approximated the stable semi-major axis for prograde and retrograde moons as,

$$a_m \leq 0.4895R_H(1.0000 - 1.0305e_p - 0.2738e_m)\ , \tag{4.77}$$

and
$$a_m \leq 0.9309 R_H (1.0000 - 1.0764 e_p - 0.9812 e_m) \, , \tag{4.78}$$

respectively. Grishin et al. (2017) generalize Hill-Stability Criteria by taking into account Lidov-Kozai mechanism and the effect of evection resonance.

However, the Hill sphere is only an approximation, and other forces can eventually perturb an object out of the sphere. Therefore, in order to investigate habitable exomoons, the moon's have to be stable long enough for life to form. The moon's orbit slowly changes due to perturbations caused by the tidal bulge of the planet. On a moon with mass M_m and semi-major axis a_m, the tidal bulge causes a torque

$$\tau_{p-m} = \frac{3}{2} \frac{k_{2p} G M_m^2 R_p^5}{Q_p a_m^6} \mathrm{sgn}(n_p - n_m) \, , \tag{4.79}$$

where G is gravitational constant, R_p is planet radius, k_{2p} is the tidal Love number of the planet, Q_p is tidal dissipation parameter, sgn is the Signum function, n_p is the angular velocity of the planet's rotation and n_m is the angular velocity of the moon's orbit (Murray and Dermott 1999). For a Jupiter-like system, Q_p and k_{2p} are around 10^5 and 0.5 respectively (Barnes and O'Brien 2002; Kipping et al. 2009).

Additionally, the total lifetime of a moon, T_{p-m} is defined by the time necessary for the moon orbit in the region between the critical semi-major axis (a_{crit}, the location of the outermost satellite orbit that remains bound to the planet) and the planet's radius (R_p) (Barnes and O'Brien 2002):

$$T_{p-m} = \frac{2}{13} (a_{crit}^{13/2} - R_p^{13/2}) \frac{Q_p}{3 k_{2p} M_m R_p^5} \sqrt{\frac{M_p}{G}} \, . \tag{4.80}$$

In a recent study, Porter and Grundy (2011) used the Kozai Cycle and Tidal Friction model to simulate captured exomoons around giants and found that exomoons could stabilise in a few million years, which is very short relative to the life time of the stars.

The transit technique is most sensitive to hot-Jupiters. Barnes and O'Brien (2002) proposed that habitable exomoons should be Earth-like mass moons orbiting around habitable-zone Jupiter-like planets around M-dwarf stars. Therefore, in this work, we will focus on habitable exomoons orbiting a hot-Jupiter around a M-dwarf star.

4.5 Measuring Exomoon Detectability

4.5.1 Kepler Transiting Light Curve Generation

In order to generate transit light curves of a planet with a moon, the algorithms of Kipping (2011a) are used. The *Kepler* mission is designed to monitor $\sim$150,000 stars brighter than 16th magnitude (in the *Kepler* passband) with 20 parts per million photometric precision at 12th magnitude in 6.5 h (Batalha et al. 2010; Caldwell et al. 2010b). In order to meet this requirement, the estimated photon collection rate of *Kepler* is (Borucki et al. 2005; Yee and Gaudi 2008; Kipping et al. 2009),

$$\Gamma_{ph} = 6.3 \times 10^{8-0.4(m-12)} \text{ photons/hour,} \qquad (4.81)$$

where m is the apparent magnitude. However, *Kepler* photometry is also affected by shot noise, background flux and instrumental noise. Table 4.5 summarises the properties we assume for *Kepler* photometry, including noise contributions we now discuss (Fig. 4.10).

Shot Noise

Shot noise or Poisson noise comes from the discrete nature of photons. At 12th magnitude, the largest noise component is the Poisson noise of the target (Caldwell et al. 2010b) which we simulated.

Background Flux

The background flux for *Kepler* comes from zodiacal light from the Solar System and diffuse starlight from background stars. In pre-launch prediction, the background flux is estimated at around 334 $e^-\sec^{-1}$ or 22 magnitudes per square arcsecond (Caldwell et al. 2010b). However, in real observations, the background flux varies across detectors and with orientation of the telescope. We adopted the pre-launch background flux estimate to generate the light curves in this work.

Instrumental Noise

There are two main components of instrumental noise for *Kepler*: read noise and dark current. From in-flight measurement, the read noise median value is 95 $e^-\text{read}^{-1}$ and the dark current is $0.25\ e^-\text{pixel}^{-1}\text{s}^{-1}$ which is low compared to the photons collected from the targets (Caldwell et al. 2010a). We included it in our simulation, though its affect on our results is negligible.

Table 4.5 *Kepler* photometry properties (Awiphan and Kerins 2013)

Parameter	Value
Exposure time (s)	6.02*
Plate scale (arcseconds/pixel)	3.98†
Background flux ($e^-\text{s}^{-1}$)	334†
Read Noise ($e^-\text{read}^{-1}$)	95‡
Dark Current ($e^-\text{pixel}^{-1}\text{s}^{-1}$)	0.25‡

*Van Cleve and Caldwell (2009)
‡Caldwell et al. (2010a)
†Caldwell et al. (2010b)

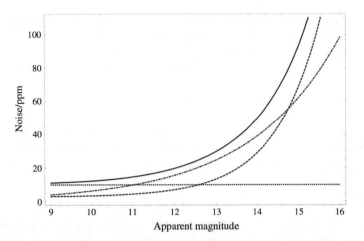

Fig. 4.10 Instrument noise (dashed), photon noise (dot-dashed), stellar variability (dotted) and total noise (solid) of *Kepler* class photometry as a function of apparent magnitude of stars (Kipping et al. 2009)

4.5.2   Measuring TTV-TDV Signals

In order to find the transit time of minimum and transit duration, the ingress and egress of our simulated light curves are fitted. The light curves are divided into phase bins using the input period which is assumed to be precisely determined from observational data. A running straight-line fit is made to three consecutive points of phased data. The fits with minimum and maximum slopes are chosen to define the ingress and egress of the transit, respectively. The intersection between the light curve median and the ingress and egress slopes are used to define the ingress (t_{ing}) and egress (t_{egr}) times, respectively (Fig. 4.11). The time of minimum light (t_0) and the transit duration ($\bar{\tau}_{mean}$) are defined as, $t_0 = (t_{ing} + t_{egr})/2$, and $\bar{\tau}_{mean} = t_{egr} - t_{ing}$, respectively.

Using the mid-transit time, a new ephemeris as a function of epoch is derived. The new ephemeris is determined by fitting a linear function to the mid-transit points.

$$T_0(n) = T_0(0) + nP , \tag{4.82}$$

where n is epoch and T_0 is time of minimum light as a function of epoch. The residuals of the times of minima and transit durations are taken as the TTV and TDV signals of the system, respectively.

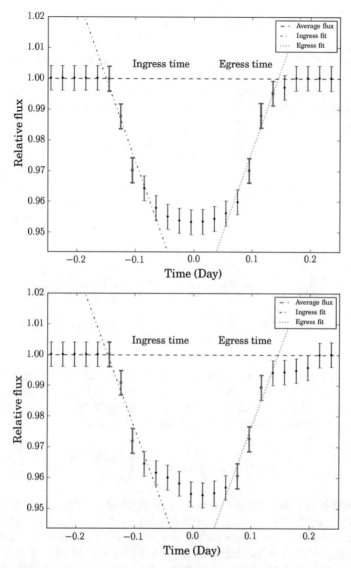

Fig. 4.11 Simulated light curves of a 15 $M_\oplus$ habitable-zone planet with a 10 $M_\oplus$ moon for an M-dwarf host star with planet period 89.35 days and moon period 2.24 days with different moon phases: Conjunction (Top) and Quadrature (Bottom). Error bars are shown at 1,000 times their true size. The fits of three red points show ingress time and egress time of the transit, and average flux (median of flux data) (Awiphan and Kerins 2013)

4.5.3   TTV-TDV Correlation Testing

From Sect. 4.3, TTV and TDV signals are sinusoidal functions and the TTV signal is 90° out of phase with the TDV signal in coplanar systems. In the case of a circular planetary orbit and a co-aligned moon orbit, the TDV-TIP component exists. From Eqs. 4.61 and 4.71, the TTV signal, TDV signal and the relation between TTV and TDV are,

$$S_{TTV} = \left[\frac{a_m M_m P_p}{2\pi a_p M_p} \right] \sin(f_m) \,, \tag{4.83}$$

$$S_{TDV} = \bar{\tau} \left[\left(\frac{a_m M_m P_p}{a_p M_p P_m} \right) + \left(\frac{b_p}{1 - b_p^2} \right) \left(\frac{a_m M_m}{R_* M_p} \right) \cos i_p \right] \cos(f_m) \,, \tag{4.84}$$

and

$$S_{TDV}^2 = -\left(\frac{2\pi a_p \bar{\tau}}{P_p} \right)^2 \left(\frac{P_p}{a_p P_m} + \frac{b_p}{1 - b_p^2} \frac{\cos i_p}{R_*} \right)^2 S_{TTV}^2$$
$$+ \left(\frac{a_m M_m \bar{\tau}}{M_p} \right)^2 \left(\frac{P_p}{a_p P_m} + \frac{b_p}{1 - b_p^2} \frac{\cos i_p}{R_*} \right)^2 \,, \tag{4.85}$$

where b_p is impact parameter of the planet from its host and R_* is host star radius. However, the ratio between TDV-V to TDV-TIP is very large. For the systems in our simulation, the minimum ratio is 1,100. Therefore, the TDV-TIP is negligible and the relation between TTV and TDV signal can be written as,

$$S_{TDV}^2 = -\left(\frac{2\pi \bar{\tau}}{P_m} \right)^2 S_{TTV}^2 + \bar{\tau}^2 \left(\frac{a_m M_m P_p}{a_p M_p P_m} \right)^2 . \tag{4.86}$$

Therefore, in theory, the plot between the square of the TTV signal and square of the TDV signal should show a perfect linear relationship with negative slope. However, there are other effects, such as star spots, instrument noise and sparsity of observation which could produce false positive TTV and TDV signatures.

In this simulation, the instrument noise and observing frequency both affect the TTV and TDV signals. Thus, the plot between S_{TTV}^2 and S_{TDV}^2 may not show a clear linear relationship. In order to check this relationship, the Pearson product-moment correlation coefficient was calculated to test the correlation between S_{TTV}^2 and S_{TDV}^2. The coefficient is

$$\chi = \frac{\sum_{i=1}^{n}(S_{TTV,i}^2 - \overline{S_{TTV}^2})(S_{TDV,i}^2 - \overline{S_{TDV}^2})}{\sqrt{\sum_{i=1}^{n}(S_{TTV,i}^2 - \overline{S_{TTV}^2})^2}\sqrt{\sum_{i=1}^{n}(S_{TDV,i}^2 - \overline{S_{TDV}^2})^2}} . \tag{4.87}$$

A negative coefficient is produced by an inverse relationship between the two variables and a positive coefficient means there is a positive linear relationship. The positive slope of S^2_{TTV} and S^2_{TDV} plot means that the TTV and TDV signal are not consistent with sinusoidal functions with a 90° phase difference. The TTV signal, TDV signal and S^2_{TTV} versus S^2_{TDV} of three sample systems are shown in Fig. 4.15. In following work, we defined correlation value to be minus-value of correlation coefficient.

4.6 Detectability of Habitable Exomoons

4.6.1 Modelling Habitable Exomoons

Properties of the Host Star

In this analysis, M-dwarf stars are selected to be the exoplanet host stars. Very cool (late K and early M type) dwarf stars have become popular targets of planet searches, because the amplitudes of the transits generated by planets in M-dwarfs are larger than those generated by hotter stars (Charbonneau et al. 2009; Bean et al. 2010b; Vogt et al. 2010; Mann et al. 2012) and the small distance of their habitable zone increases the transit probability of habitable planets as well as the transit frequency per observation time (Kaltenegger 2010). Sasaki et al. (2012) also suggested that the semimajor axis of the host planet for the most detectable exomoons around an M-dwarf star is 0.2–0.4 AU. Therefore, the most detectable exomoons in M-dwarf systems can orbit within the habitable zone. However, the host with mass less than $0.2\,M_{\oplus}$ cannot host a habitable exomoon (Heller 2012).

Kepler monitored the Cygnus region along the Orion arm centred where there are about 0.5 million stars brighter than 16th magnitude (*Kepler* passband) within its FOV. However, only 10^5 stars with magnitude brighter than 16 are expected to be exoplanet hosts. In 2010 the *Kepler* mission announced 150,000 highest priority target stars, but only 2% of these target stars have effective temperature less than 3500 K (Batalha et al. 2010), whereas >70% of all stars within 20 pc are M-dwarfs (Henry et al. 1994; Chabrier 2003; Reid et al. 2004). However, in 2011, the team released additional exoplanet data, including 997 planet-candidate host stars in which 74 (>5%) have effective temperature less than 4400 K in the *Kepler* Input Catalog (Batalha et al. 2010; Borucki et al. 2011b; Brown et al. 2011).

For our simulation of TTV and TDV signals, we assume that the host is an M-dwarf star with mass $0.5\,M_{\odot}$ and radius $0.55\,R_{\odot}$. Their effective temperature, microturbulent velocity and $\log g$ are set to be 3500 K, 1 km.s^{-1} and 4.5, respectively, as applicable to solar-metallicity M-dwarf (Bean et al. 2006; Önehag et al. 2012). In order to calculate the limb darkening coefficient, solar-metallicity is assumed and a quadratic limb-darkening model is used. The values of limb-darkening coefficients for the transmission curves of *Kepler* are obtained from Claret and Bloemen (2011).

For our M-dwarf targets, the value of the coefficients γ_1 and γ_2 are 0.4042 and 0.3268, respectively.[1]

Properties of the Host Planet

Jupiter-like giant planets offer the best potential for detecting habitable exomoons (Kipping et al. 2009). In order to investigate habitable exoplanets and exomoons, the planet-star separation is set to be inside the habitable zone, starting at a separation of 0.10 AU and increasing in logarithmically to 0.66 AU. This range includes the semimajor axis of M-dwarf planets which Sasaki et al. (2012) argue on stability ground may be among first detectable exomoon systems.

We simulate giant planets with masses ranging logarithmically from 15 to 150 $M_\oplus$. Fortney et al. (2007b) found that the radius of giant planets depends on their overall mass, core mass and separation. For giant planets of age 4.5 Gyr, their radius falls between 1.0 and 1.2 R_J (Jupiter radius). Therefore, we adopt a planet radius of 1.2 R_J.

4.6.1.1 Properties of the Exomoon

No exomoon has yet been discovered, therefore the properties of Earth-like planets are used for the habitable exomoon in this work. Rocky planets with logarithmic mass between 1 and 10 $M_\oplus$ are chosen. The radius of the moon is calculated from Fortney's model, using rock mass fraction equal to 0.66 (Earth-like planet):

$$R_m = 1.00 + 0.65 \log M_m + 0.14 (\log M_m)^2 \tag{4.88}$$

where M_m and R_m are the moon's mass and moon's radius in $M_\oplus$ and $R_\oplus$, respectively (Fortney et al. 2007a, b). Only a moon within the planet's Hill sphere with an orbital period between 1.00 and 3.16 days is considered. Again, for simplicity, circular orbits are assumed.

4.6.2 Detectability of Habitable Exomoons

The light curves are generated with *Kepler* photometric noise. 146,410 light curves are simulated with 11 independent values of each of four variable input parameters: planet mass; planet separation; moon mass; and moon period, and 10 random initial phases. The host stars are assumed to be M-dwarf stars of 12.5 magnitude in the *Kepler* passband. The cadence of this simulation is 50 data points per day (every 28.8 mins) which corresponds closely to *Kepler*'s long cadence mode (every 29.4 mins) (Gilliland et al. 2010). In order to simulate the current *Kepler* data, a 3-year

---

[1] See http://cdsarc.u-strasbg.fr/viz-bin/qcat?J/A+A/529/A75.

Table 4.6 Input parameters assumed for our exomoon simulations (Awiphan and Kerins 2013)

Star parameters	
Mass ($M_\odot$)	0.5
Radius ($R_\odot$)	0.55
Apparent magnitude (K_p)	12.5
Quadratic limb-darkening coefficient 1	0.4042
Quadratic limb-darkening coefficient 2	0.3268
Planet parameters	
Mass ($M_\oplus$)	15.0–150.0
Radius (R_J)	1.2
Separation (AU)	0.10–0.66
Eccentricity	0.0
Inclination (degrees)	90.0
Moon parameters	
Mass ($M_\oplus$)	1.0–10.0
Radius ($R_\oplus$)	Equation 4.88
Period (days)	1.00–3.16
Eccentricity	0.0
Inclination (degrees)	90.0

simulation of a transiting giant extrasolar planet with a rocky extrasolar moon was run to find out the detectability of an exomoon in the M-dwarf habitable zone. The details of physical parameters of the systems are listed in Table 4.6.

The 4D-simulation is projected on to two-parameter planes in order to examine the relation between two variables. Since we are only interested in negative correlations, we define the projected correlation as:

$$\chi_{\text{proj}} = \frac{1}{N} \sum_{i(\chi>0)}^{N} \chi_i , \qquad (4.89)$$

where N is the total number of 2-D simulations that are projected and $i(\chi > 0)$ refers only to those simulations with negative correlation. The projected plots therefore represent averages over logarithmic parameter priors for negative correlation signals. In the left hand panels of Fig. 4.12a, the plot between planet mass and moon mass shows that a high-mass moon hosted by a low-mass planet is the most detectable of the systems considered. This result agrees with the moon period versus planet mass and moon mass plots (Fig. 4.12e, f). However, in these two plots, the changes in moon period do not affect the correlation. In Fig. 4.12b, the projection plot between separation of planet and period of moon also does not show any significant trends.

Figure 4.12c shows the detectability coefficient between mass and separation of the planet. Planets with large separation have higher detectability than close-in planets of the same mass. This result correlates with the result of moon mass versus planet separation (Fig. 4.12d) which shows that, in systems of equal satellite mass, the outer planet-moon systems have larger correlation coefficients. These features may be produced by only a few transit events in systems with large star-planet separation, because, at 0.6 AU separations, only three transit events are detected in the simulation. Therefore, we now check the reliability of the correlation.

The analysis of correlations is meaningful when the correlations are not dominated by noise. The variance of correlation is plotted in the right-hand panels of Fig. 4.12 in order to check the reliability of testing. From Fig. 4.12a–d, the variance plots show that the systems with a small number of transit events (planets with long orbital period) have higher variance. However, the value of the variance is still low compared to the correlation coefficient.

While the magnitude of χ_{proj} in Eq. 4.89 is reduced by positive correlations that are included in N, we have checked that the basic features in the plots of Fig. 4.12 trace those obtained by ignoring positive correlations, albeit at a weaker level. Finally, the assumption that moons of outer planets should be easier to detect than moons of inner planets is confirmed by our simulations.

The theoretical lines of RMS amplitude of the TTV and TDV signals are shown to investigate the features of the contours. For Fig. 4.12a the high amplitude of TTV and TDV signals produces a high coefficient of detection with the same slope. Moreover, the features in Fig. 4.12c, d are also well-correlated with TDV RMS amplitude signals which can be explained by the relative weakness of TDV signals compared with TTV signals. In conclusion, the detectability of exomoons is dominated by the amplitude of TDV signals.

The structures of the correlation plots are explained by the magnitude of the TDV signal. However, in Fig. 4.12c, d, gaps are evident at planet semi-major axes of 0.4 and 0.5 AU. The variance plots show that there is no difference in variance across this region and that therefore the features in Fig. 4.12c, d are real. To investigate the structures, the 4-D plots are sliced into 2-D plots. The correlation plots of planet separation versus planet mass, and of planet separation versus moon mass are shown in Figs. 4.13 and 4.14. Systems with high moon mass and low planet mass have a high value of correlation and the features in these contours correspond to the projected contours of Fig. 4.12, including the gap structures.

In Figs. 4.13 and 4.14, the maps with similar moon period show gap features at the same planet separation, but they shift with a different moon period. The ratios between moon period and planet which produce the gap are near-integer values and correspond to the cold spots in Fig. 4.12b. Therefore, the gap structures can be explained by a resonance between the planet and moon period which produces constant detected TTV and TDV signals. However, they also depend on the number of detected transit events. In short period systems which have a larger number of

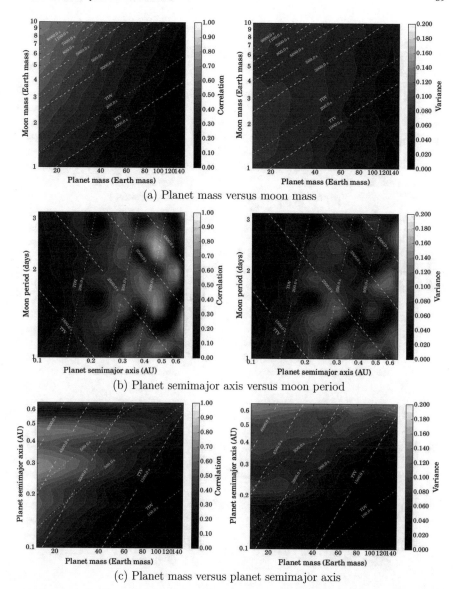

(a) Planet mass versus moon mass

(b) Planet semimajor axis versus moon period

(c) Planet mass versus planet semimajor axis

Fig. 4.12 Correlation (left) and variance of correlation (right) between **a** planet mass and moon mass, **b** planet semimajor axis and moon period, **c** planet mass and planet semimajor axis, **d** moon mass and planet semimajor axis, **e** planet mass versus moon period, and **f** moon mass and moon period of the light curves. The contour is averaged over other two variable. The RMS amplitude of the TTV signal (dashed) and RMS amplitude of the TDV signal (dot-dashed) in units of seconds are presented. The cold spots in **b** indicate the data with the planet period in resonance with the moon period (Awiphan and Kerins 2013)

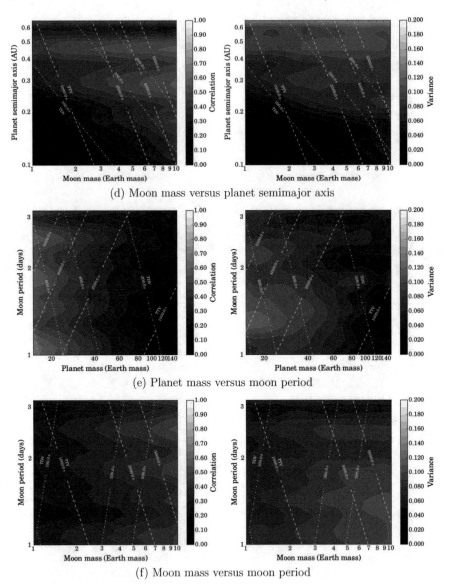

(d) Moon mass versus planet semimajor axis

(e) Planet mass versus moon period

(f) Moon mass versus moon period

Fig. 4.12 (continued)

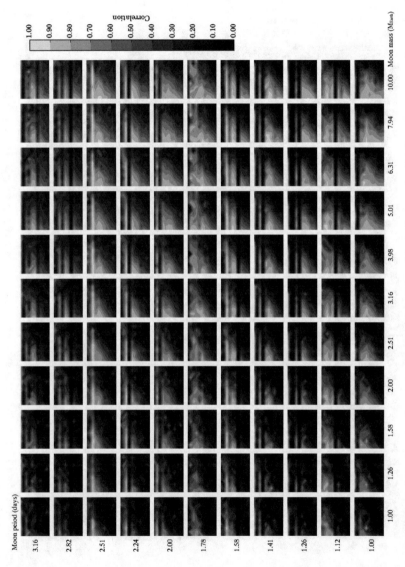

Fig. 4.13 The correlation plots of planet separation versus planet mass on a logarithmically spaced grid. The axis ranges are the same as the range used in Fig. 4.12c (Awiphan and Kerins 2013)

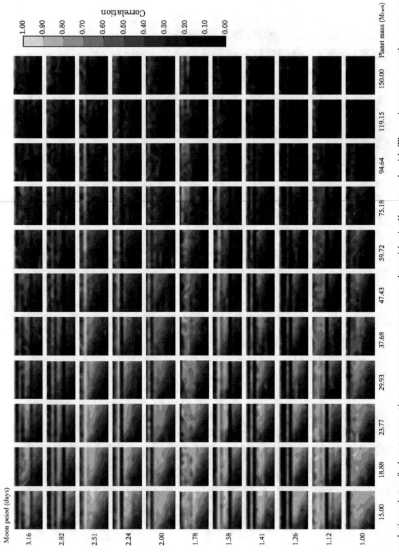

Fig. 4.14 The correlation plots of planet separation versus moon mass on a logarithmically spaced grid. The axis ranges are the same as the range used in Fig. 4.12d (Awiphan and Kerins 2013)

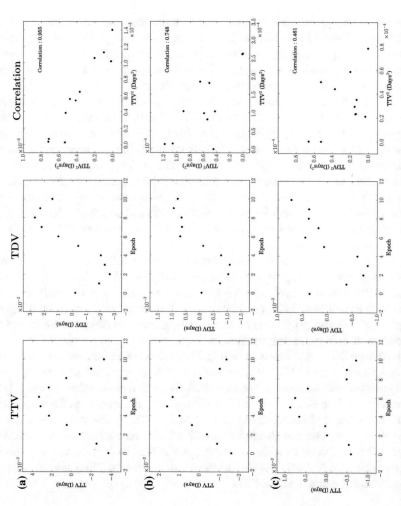

Fig. 4.15 TTV signal (left), TDV signal (middle) and S_{TTV}^2 versus S_{TDV}^2 (right) for a 10.0 $M_\oplus$ exomoon of a planet orbiting around a 0.5 $M_\odot$ M-dwarf star. The planet period is 89.35 days and moon period is 2.24 days. Panels are shown for three exoplanets with $\log(M_P/15M_\oplus)$ equal to 0.0 (**a**), 0.4 (**b**), and 0.7 (**c**) (Awiphan and Kerins 2013)

transits, the gap structures are more difficult to produce due to the larger range of detected planetary phases (Fig. 4.15).

4.7   Exomoon Detectability with Interstellar Noise

In reality, stars can vary in brightness due to pulsation, rotation and activity (red noise). Around 40–70% of M-dwarfs have variability with photometric dispersion (σ_m) $\sim$3–5 mmag, depending on their brightness. Normally, the error in the ingress and egress parts dominates the uncertainties in the mid-transit time and transit duration. However, the red noise also affects the uncertainties over longer period.

For stars with 12.5 magnitude in the *Kepler* passband, the variability fraction with amplitude $\sigma > 0.1$ is nearly 1 (Ciardi et al. 2011) and their noise tends to have long variation periods ($\geq$5 days) (McQuillan et al. 2012). Therefore, in our simulation, the red noise with a 12 days period was added into the light curves in order to investigate the effect of stellar noise to the detectability. The stellar variability amplitude is based on that found for M-dwarfs with *Kepler* magnitude between 12 and 14 (Ciardi et al. 2011). In order to simulate short term variability, five minor variations with uniformly distributed periods between 0 and 12 days and amplitudes less than half of the amplitude of the main variation were also added. An example of our simulated red noise is shown in Fig. 4.16.

Our test of the effects of red noise was based on the following parameters. We simulated planets with $\log(M_p/15M_\oplus)$ equal to 0.0, 0.1, 0.2, 0.3, 0.4 and 0.5 and tested moons with $\log(M_m/M_\oplus)$ equal to 0.8, 0.9 and 1.0. The planet and moon are assumed to have a period of around 89 and 2.2 days, respectively, corresponding to the peak. The stellar parameters of M-dwarf host stars are described in Table 4.6. The values of limb-darkening coefficients for the transmission curves of *Kepler* are obtained from Claret and Bloemen (2011). For each set of parameters used, we used simulations with high TTV-TDV correlation ($>$0.7) to examine the effect of red noise on high-confidence detections. We simulated 500 different variations to each light curve. The result in Fig. 4.17 shows that the presence of intrinsic stellar variability of M-dwarfs might affect the exomoon detectability. The stellar variability reduces the exomoon detection correlation by 0.0–0.2 with 0.1 median reduction. However, for our simulated systems with planet masses less than around 25 $M_\oplus$ with moon masses 8–10 $M_\oplus$, typically 25–50% of them still have correlations high enough to be confirmed as exomoon detections.

4.8   Conclusion

The light curves of a transiting exoplanet with an exomoon were simulated for the purpose of determining the detectability of exomoons. The *Kepler* photometric noise was modelled to the light curve in order to simulate the data from *Kepler*. Quantifying

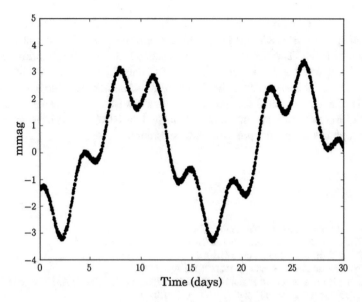

Fig. 4.16 Our simulated stellar red noise with a main noise component of 12 days period (Awiphan and Kerins 2013)

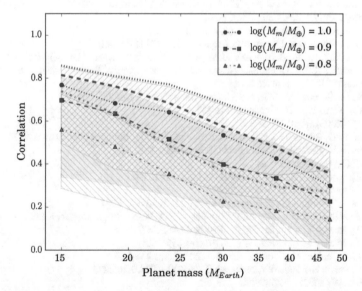

Fig. 4.17 Median correlation as a function of planet mass. We assume an exomoon with $\log(M_p/15M_\oplus)$ equal to 1.0 (Blue circle with dot line), 0.9 (Red square with dashed line) and 0.8 (Green triangle with dashed-dot line) orbiting around an M-dwarf with planet period and moon period of 89.35 and 2.24 days, respectively. Thick lines show the median correlations of the systems without stellar variability. The forward diagonal hatch region, shaded area and backward diagonal hatch region represents a 25–75th percentile region of systems with 1.0, 0.9 and 0.8 $M_\oplus$ planets, respectively, in the presence of red noise (Awiphan and Kerins 2013)

the detectability was done by phase-correlation between TTV and TDV signals. TTV and TDV always exhibit a $90°$ phase shift. Therefore, the TTV^2 signal is linearly correlated with the TDV^2 signal. The Pearson product-moment correlation coefficient was used to determine the detectability of the signals. The effects of intrinsic stellar variation of an M-dwarf reduce the detectability correlation coefficient by 0.0–0.2 with 0.1 median reduction. For simulations with red noise with planet masses less than around 25 $M_\oplus$, 25–50% of simulated systems with 8–10 $M_\oplus$ moon have correlations high enough to confirm the presence of an exomoon.

References

Awiphan S, Kerins E (2013) MNRAS 432:2549
Baines EK, van Belle GT, ten Brummelaar TA et al (2007) ApJ 661:L195
Barnes JW, O'Brien DP (2002) ApJ 575:1087
Batalha NM, Borucki WJ, Koch DG et al (2010) ApJ 713:L109
Bean JL, Sneden C, Hauschildt PH, Johns-Krull CM, Benedict GF (2006) ApJ 652:1604
Bean JL, Seifahrt A, Hartman H et al (2010b) ApJ 713:410
Ben-Jaffel L, Ballester GE (2014) ApJ 785:L30
Bennett DP, Batista V, Bond IA et al (2014) ApJ 785:155
Borucki WJ, Koch D, Basri G et al (2005) in A Decade of Extrasolar Planets around Normal Stars.
    Cambridge Univ. Press, Cambridge, p 36
Borucki WJ, Koch DG, Basri G et al (2011b) ApJ 736:19
Brown TM, Latham DW, Everett ME, Esquerdo GA (2011) AJ 142: 112
Caldwell D. A., van Cleve J. E., Jenkins J. M., et al., 2010b, in Society of Photo-Optical Instrumenta-
    tion Engineers (SPIE) Conference Series, vol. 7731 of Society of Photo-Optical Instrumentation
    Engineers (SPIE) Conference Series
Caldwell DA, Kolodziejczak JJ, Van Cleve JE et al (2010a) ApJ 713:L92
Canup RM, Ward WR (2006) Nature 441:834
Chabrier G (2003) PASP 115:763
Charbonneau D, Brown TM, Burrows A, Laughlin G (2007) Protostars and planets V, 701–716
Charbonneau D, Berta ZK, Irwin J et al (2009) Nature 462:891
Chyba CF (1997) Nature 385:201
Ciardi DR, von Braun K, Bryden G et al (2011) AJ 141: 108
Claret A, Bloemen S (2011) A&A 529:A75
Domingos RC, Winter OC, Yokoyama T (2006) MNRAS 373:1227
Fewell M (2006) Technical Report, DSTO-TN- 0722 (Online). http://hdl.handle.net/1947/4551
Fortney JJ, Marley MS, Barnes JW (2007a) ApJ 668:1267
Fortney JJ, Marley MS, Barnes JW (2007b) ApJ 659:1661
Gilliland RL, Jenkins JM, Borucki WJ et al (2010) ApJ 713:L160
Grishin E, Perets HB, Zenati Y, Michaely E (2017) MNRAS 466:276
Han C, Han W (2002) ApJ 580:490
Heller R (2012) A&A 545:L8
Heller R (2014) ApJ 787:14
Heller R, Williams D, Kipping D et al (2014) Astrobiology 14:798
Henry TJ, Kirkpatrick JD, Simons DA (1994) AJ 108: 1437
Jordán A, Bakos GÁ (2008) ApJ 685:543
Kaltenegger L (2010) ApJ 712:L125
Kipping DM (2009a) MNRAS 392:181
Kipping DM (2009b) MNRAS 396:1797

Kipping DM (2011a) MNRAS 416:689
Kipping DM (2011b) The transits of extrasolar planets with moons. Springer, Berlin
Kipping DM, Fossey SJ, Campanella G (2009) MNRAS 400:398
Laskar J, Joutel F, Robutel P (1993) Nature 361:615
Lewis KM, Sackett PD, Mardling RA (2008) ApJ 685:L153
Liebig C, Wambsganss J (2010) A&A 520:A68
Mandel K, Agol E (2002) ApJ 580:L171
Mann AW, Gaidos E, Lépine S, Hilton EJ (2012) ApJ 753:90
McLaughlin DB (1924) ApJ 60
McQuillan A, Aigrain S, Roberts S (2012) A&A 539:A137
Murray CD, Dermott SF (1999) Solar system dynamics. Cambridge University Press, New York
Noyola JP, Satyal S, Musielak ZE (2014) ApJ 791:25
Önehag A, Heiter U, Gustafsson B, Piskunov N, Plez B, Reiners A (2012) A&A 542:A33
Porter SB, Grundy WM (2011) ApJ 736:L14
Rafikov RR (2009) ApJ 700:965
Reid IN, Cruz KL, Allen P et al (2004) AJ **128**: 463
Rossiter RA (1924) ApJ 60
Sartoretti P, Schneider J (1999) A&AS 134:553
Sasaki T, Barnes JW, O'Brien DP (2012) ApJ 754:51
Scharf CA (2007) ApJ 661:1218
Simon A, Szatmáry K, Szabó GM (2007) A&A 470:727
Simon AE, Szabó GM, Szatmáry K, Kiss LL (2010) MNRAS 406:2038
Simon AE, Szabó GM, Kiss LL, Szatmáry K (2012) MNRAS 419:164
Sumi T (2010) In Pathways towards habitable planets, ed. by Coudé Du Foresto V, Gelino DM, Ribas I, vol. 430 of Astronomical society of the pacific conference series, 225
Szabó GM, Szatmáry K, Divéki Z, Simon A (2006) A&A 450:395
Teachey A, Kipping DM, Schmitt AR (2018) AJ 155:36
Van Cleve JE, Caldwell DA (2009) Kepler instrument handbook, KSCI 19033-001. NASA Ames Research Center
Vogt SS, Butler RP, Rivera EJ, Haghighipour N, Henry GW, Williamson MH (2010) ApJ 723:954
Williams DM, Knacke RF (2004) Astrobiology 4:400
Williams DM, Kasting JF, Wade RA (1997) Nature 385:234
Yee JC, Gaudi BS (2008) ApJ 688:616
Zhuang Q, Gao X, Yu Q (2012) ApJ 758:111

Chapter 5
Microlensing Maps of the Galaxy

For longer than a decade, several microlensing surveys have monitored a large number of stars and detected thousands of microlensing events over the Galactic bulge (Hamadache et al. 2006; Popowski et al. 2005; Sumi et al. 2003, 2006, 2013; Wyrzykowski et al. 2015; Mróz et al. 2017). The microlensing optical depth, average time scale and microlensing rate are key parameters which provide basic information of microlensing events. The theoretical prediction maps of these parameters are very useful for survey area selection for both on-going and future microlensing surveys.

Kerins et al. (2009) presented synthetic maps of these parameters over the Galactic bulge using catalogues generated from the Besançon Galactic model (Robin et al. 2003; Marshall et al. 2006). However, the simulated maps are available in only I, J, and K bands. In this work, the latest version of Besançon Galactic model is used to simulate the first real-time web-based microlensing map. We intend to add some additional tools to constrain the parameters of the lens and source, such as filter, source magnitude limit and relative proper motion, to the map, which will be useful for the future microlensing surveys toward the Galactic bulge.

5.1 Gravitational Microlensing

5.1.1 Gravitational Microlensing Background

The gravitational microlensing technique was first proposed by Einstein (1936); Liebes (1964) and Refsdal (1964). It occurs when a foreground lens object passes close to an observer's line of sight to a background source. The gravitational field of the lensing object will perturb the light rays, which creates multiple images, and magnifies or demagnifies the source flux. Moreover, the presence of additional objects in the lensing system, such as stellar or planetary companions, will create an additional perturbation

© Springer International Publishing AG, part of Springer Nature 2018
S. Awiphan, *Exomoons to Galactic Structure*, Springer Theses,
https://doi.org/10.1007/978-3-319-90957-8_5

to the brightness of the source. However, microlensing events are rare, unpredictable and unrepeatable (Gaudi 2012).

The microlensing surveys toward the Galactic bulge have provided useful information for the search for exoplanets and for the study of Galactic structure (see Sect. 5.1.2 for the study of Galactic structure, Paczynski 1996 and Gaudi 2012). For exoplanet detection, the microlensing technique is sensitive to systems at large distances from the Earth and for a wide range of lens masses, including low-mass planets, wide-separation planets, free-floating planets and planets beyond the snow line, because it does not require the detection of flux from the planet or the host star (Kennedy and Kenyon 2008; Lecar et al. 2006). Therefore, microlensing can provide the demographics of planetary systems throughout the Galaxy without bias towards bright or nearby stars. However, in order to fully characterize the properties of large numbers of distant and faint systems, space-based microlensing missions that can isolate host star light from the background stars in crowded fields are required (Gaudi 2012).

The first planet that was discovered using microlensing was OGLE 2003-BLG-235/MOA 2003-BLG-53 in 2004 (Bond et al. 2004). Currently, more than 40 planets have been detected using this method. In the future, the number of planet detections using microlensing will increase due to the ongoing main microlensing surveys, MOA (Bond et al. 2002), OGLE (Udalski et al. 2008) and KMTNet (Henderson et al. 2014), and also future space-based microlensing surveys, such as WFIRST (Spergel et al. 2013, 2015) and Euclid (Laureijs et al. 2011).

5.1.2   Single Lens Microlensing

The basic geometry of microlensing of point mass lenses is shown in Fig. 5.1. A single point mass M (a lens) is located at a distance D_l, and background source is located at distance D_s. When the source passes by the lens with angular separation β, the source light is deflected by the gravity of the lens by an angle $\hat{\alpha}_d$. The images of the source with the apparent angular position θ are displaced by an angle $\alpha_d = \theta - \beta$,

$$\beta = \theta - \alpha_d = \theta - \left(\frac{D_s - D_l}{D_s} \right) \hat{\alpha}_d . \tag{5.1}$$

From General Relativity, the deflection angle of a light ray in the gravitational field of a point mass is

$$\hat{\alpha}_d = \frac{4GM}{c^2} \frac{b}{|b|^2} , \tag{5.2}$$

where G is the gravitational constant, c is the speed of light and b is the distance vector between the light ray and the lens (Einstein 1915). If the lens and source are exactly aligned ($\beta = 0$), the source is imaged into a ring of angular radius

Fig. 5.1 The geometry of light paths of a microlensing event. Light leaves the source S, passes the lens L and reaches the observer O, appearing as two images (I_+ and I_-)

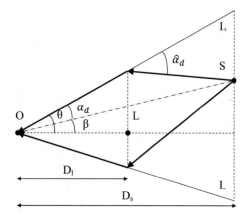

$$\theta_E = \sqrt{\frac{4GM(D_s - D_l)}{c^2 D_s D_l}} \tag{5.3}$$

called the Einstein radius, which is the characteristic angular separation for gravitational lensing. For the lensing effect of stars on other stars in the Galaxy, the Einstein radius can be written as,

$$\theta_E = 2.85\text{mas} \left(\frac{M}{M_\odot}\right)^{1/2} \left(\frac{1-x}{x}\right)^{1/2} \left(\frac{D_s}{\text{kpc}}\right)^{1/2}, \tag{5.4}$$

where $x = D_l/D_s$. The physical Einstein radius in the plane of the lens is

$$R_E = \theta_E D_l. \tag{5.5}$$

Therefore, the single lens equation (Eq. 5.1) can be written in simple form as,

$$u = y - \frac{1}{y}, \tag{5.6}$$

where $u = \beta/\theta_E$ is the impact parameter and $y = \theta/\theta_E$, which has two solutions:

$$y_\pm = \pm\frac{1}{2}(\sqrt{u^2 + 4} \pm u). \tag{5.7}$$

The positive solution (the major image) is located outside the Einstein radius ($y_+ > 1$) and the negative solution (the minor image) is located inside the Einstein ring ($|y_-| < 1$). For microlensing events in the Galaxy, the angular separation between the two images cannot be resolved by optical telescopes ($\Delta\theta = |y_+ - y_-|\theta_E = (u^2 + 4)^{1/2}\theta_E \lesssim 2\theta_E$, where $u \lesssim 1$). The magnification of the source in polar coordinates (u, ϕ) is obtained by examining the geometry of the images.

The magnification of each image can be written as,

$$A_{\pm} = \frac{y_{\pm}}{u} \left| \frac{dy_{\pm}}{du} \right| ,$$

$$= \frac{1}{2} \left(\frac{u^2 + 2}{u\sqrt{u^2 + 4}} \pm 1 \right) . \tag{5.8}$$

Therefore, the total absolute magnification is,

$$A = A_+ + A_- ,$$

$$= \frac{u^2 + 2}{u\sqrt{u^2 + 4}} . \tag{5.9}$$

The source, lens and observer are all in relative motion. Therefore, the angular separation between the source and lens is a function of time. For a system with uniform relative proper motion, μ, the impact parameter change with time, $u(t)$, can be written as

$$u(t) = \left[u_0^2 + \left(\frac{t - t_0}{t_E} \right)^2 \right]^{1/2} , \tag{5.10}$$

where u_0 is the minimum impact parameter of the event, t_0 is the time of closest alignment ($u = u_0$) and t_E is the Einstein radius crossing time,

$$t_E = \frac{\theta_E}{\mu} . \tag{5.11}$$

Figure 5.2 shows examples of microlensing light curves with different values of u_0.

5.2 Microlensing Properties

5.2.1 Optical Depth

Microlensing events are unpredictable. In order to calculate the probability of microlensing along a line of sight at any given time, the optical depth which is the number of ongoing events at a given time, is used. Microlensing events are usually defined as events having an angular separation between source and lens not more than the angular Einstein ring radius of the lens. Therefore, the optical depth is the fraction of sky that is covered by the Einstein rings of foreground lenses within same solid angle, Ω. For a shell with solid angle Ω at a distance D_l with thickness dD_l, the number of lenses in the shell is $n(D_l)\Omega D_l^2 dD_l$, where n is the number density of lenses. Therefore, the optical depth can be written as,

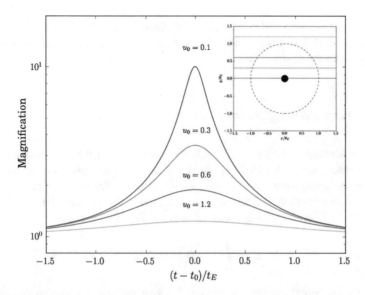

Fig. 5.2 Microlensing light curve with different values of minimum impact parameter

$$\tau = \int_0^{D_s} n(D_l) D_l^2 \pi \theta_E^2 dD_l \ . \tag{5.12}$$

The probability, P, that sources are microlensed is

$$P = 1 - e^{-\tau} \ . \tag{5.13}$$

In the case of Galactic microlensing, the optical depth is small. Then, the probability can be written as $P \simeq \tau$

Optical depth directly relates to the mass density of compact objects along the line of sight as the mass density of lenses of mass M_l is $\rho = nM_l$ and $\pi \theta_E^2 \propto M_l$. Therefore, the optical depth depends only on the mass density along the light of sight; it is independent of both lens mass and kinematics.

5.2.2 Average Einstein Radius Crossing Time

The average Einstein radius crossing time is

$$\langle t_E \rangle = \frac{\sum^{N_{ev}} t_E}{N_{ev}} \ . \tag{5.14}$$

where N_{ev} is the number of events and t_E is defined by Eq. 5.11. Although, t_E depends on the lens mass, lens distance and velocity, the t_E distribution can be used to probe the mass function of the lens objects.

5.2.3   Microlensing Event Rate

The microlesning event rate is the rate at which a background source undergoes a microlensing event due to foreground lenses. Assuming events occur when the lens-source angular separation comes within Einstein radius, the probability that microlensing will occur in a time dt is the fraction of the sky covered by solid angle $2\theta_E$ and μdt. Therefore, the microlensing event rate of an event is

$$\Gamma = \int_0^{D_s} n(D_l)\mu D_l^2 2\theta_E dD_l \ . \tag{5.15}$$

As relative proper motion is related to the crossing time, for a given line of sight, the total microlensing rate is

$$\Gamma = \frac{2\tau}{\pi \langle t_E \rangle} \ , \tag{5.16}$$

where $\langle t_E \rangle$ is the event timescale averaged over all events.

5.3   Microlensing Towards the Galactic Bulge

The microlensing surveys toward the Galactic bulge have provided useful information for the search for exoplanets and for the study of Galactic structure (Paczynski 1996; Gaudi 2012). Several microlensing surveys have monitored a large number of stars and detected thousands of events over the bulge [e.g. OGLE (Udalski et al. 1994; Sumi et al. 2006; Wyrzykowski et al. 2015), MOA (Sumi et al. 2003, 2013), KMTNet (Hwang et al. 2015), MACHO (Alcock et al. 1997, 2000; Popowski et al. 2005) and EROS (Afonso et al. 2003; Hamadache et al. 2006)]. The microlensing optical depth, τ, measures the fraction of the sky covered by the Einstein rings of the lenses for a given line of sight. As the optical depth is directly related to the mass density of the lens population, it can be used to determine the mass distribution of the bulge. However, a difficulty in measuring the microlensing optical depth stems from the fact that it is sensitive to the individual contributions of long duration events. Another measurable property from the surveys is the microlensing event rate, Γ, which has the advantage that it is not dominated by a small number of long duration events but the disadvantage that it is sensitive to Galactic kinematics and the stellar mass function, as well as the mass distribution.

A number of measurements of the bulge optical depth have been made by the survey teams, often under different sample definitions. We loosely categorize these as: resolved source measurements, difference image analysis (DIA) source measurements and red clump giant (RCG) source measurements. The resolved source method includes all sources which are brighter than the survey magnitude limit, whilst the DIA method includes fainter sources which may only brighter above the survey limit during lensing. The DIA method has the benefit that it is less sensitive to systematic blending within crowded fields and potentially provides a better S/N ratio measurement due to the larger available sample size. At the other extreme the RCG method uses samples of events which involve only bright sources that are assumed to be well resolved and therefore should exhibit a minimal blending bias. In recent studies, DIA optical depth measurements tend to be about 25% higher that those derived from RCG samples (Sumi et al. 2013).

The MOA-II survey (Sumi et al. 2013) determined the optical depth from a study of 474 events with sources brighter than 20th magnitude in the I-band toward the bulge. They determined a value of $\tau_{\mathrm{DIA}} = [2.35 \pm 0.18]e^{[0.51 \pm 0.07](3-|b|)} \times 10^{-6}$ [1]. For the average optical depth, they found that $\tau_{\mathrm{DIA},200} = 3.64^{+0.51}_{-0.45} \times 10^{-6}$ at ($l = 0.97°$, $b = -2.26°$). These results are broadly consistent with previous measurements from MOA-I (Sumi et al. 2003), OGLE (Sumi et al. 2006), MACHO (Popowski et al. 2005) and EROS Hamadache et al. (2006) (Table 5.1 and Sect. 6.1).

In the recent years, more detailed theoretical models have been developed in order to predict the microlensing optical depth values (Han and Gould 2003; Wood and Mao 2005; Kerins et al. 2009). Kerins et al. (2009) presented synthetic maps of optical depth and event rate over the Galactic bulge using catalogues generated from the Besançon Galactic model developed by Robin et al. (2003) with 3D extinction maps from Marshall et al. (2006). The observational result tends to agree with the theoretical models. However, the recent MOA-II surveys provide an optical depth for RCG sources that is 30–40% below the prediction of Kerins et al. (2009), which might be the result of lacking long crossing time events in the observational data or could indicate deficiencies in the model (Sumi et al. 2013).

5.4 The Besaçon Galactic Model

The Besançon model, a Galactic population synthesis model, is designed to describe the observable properties of the Galactic stellar population by relating them to models of Galactic formation and evolution, stellar formation and evolution and stellar atmospheres in any given line of sight, using constraints from observational data (Robin et al. 2003, 2012, 2014). In the Besançon model, stars are created from gas following an initial mass function (IMF) and star formation rate (SFR), and evolved according to theoretical stellar evolutionary tracks. For each simulated star, the photometry, kinematics and metallicity are computed. In order to simulate the Galaxy,

---

[1] The subscripts of optical depths indicate the method of analysis and the long duration cutoff in days.

Table 5.1 Observed microlensing optical depth and rate measurements towards the Galactic Centre (Awiphan et al. 2016b)

Project	Field deg²	Method	$t_{E,max}$ days	N_{event} events	l^*, b^*	τ ×10⁶	Γ_{deg^2} yr⁻¹deg⁻²	Γ_{star} ×10⁻⁶yr⁻¹star⁻¹
OGLE(1994)[1]	0.81	Resolved	100	9	$\pm 5°, -3.5°$	3.3 ± 1.2	–	–
MACHO(1997)[2]	12.0	Resolved	150	45	$2.55°, 3.64°$	$3.9^{+1.8}_{-1.2}$	–	–
MACHO(2000)[3]	4.0	DIA	150	99	$2.68°, -3.35°$	$2.43^{+0.39}_{-0.38}$	–	–
EROS(2003)[4]	15.0	RCG	400	16	$2.5°, -4.0°$	0.94 ± 0.29	–	–
MOA(2003)[5]	18.0	DIA	150	28	$3.0°, -3.8°$	$2.59^{+0.84}_{-0.64}$	–	–
MACHO(2005)[6]	4.5	RCG	350	62	$1.5°, -2.68°^{c}$	$2.17^{+0.47}_{-0.38}$	–	–
OGLE(2006)[7]	5.0	RCG	400	32	$1.16°, -2.75°$	$2.55^{+0.57}_{-0.46}$	–	–
EROS(2006)[8]	66.0	RCG	400	25	$(-6.00°·10\,00°), 1.75°$	3.52 ± 1.00	–	–
				22	$(-6.00°, 10.00°), 2.26°$	2.38 ± 0.72	–	–
				24	$(-6.00°, 10.00°), 2.76°$	1.31 ± 0.38	–	–
				25	$(-6.00°, 10.00°), 3.23°$	2.21 ± 0.62	–	–
				24	$(-6.00°, 10.00°), 4.45°$	0.92 ± 0.72	–	–
MOA(2013)[9]	3.2	DIA	200	12	$(-5.00°, 5.00°), -1.40°$	$4.47^{+1.69}_{-1.21}$	$62.4^{+22.1}_{-16.3}$	$71.2^{+25.2}_{-18.6}$
				52	$(-5.00°, 5.00°), -1.77°$	$5.01^{+1.12}_{-0.91}$	$90.9^{+14.5}_{-12.6}$	$48.2^{+7.7}_{-6.7}$
				70	$(-5.00°, 5.00°), -2.26°$	$3.49^{+0.81}_{-0.66}$	$88.6^{+13.7}_{-11.7}$	$41.1^{+6.4}_{-5.4}$
				75	$(-5.00°, 5.00°), -2.76°$	$3.33^{+0.88}_{-0.69}$	$68.8^{+9.0}_{-7.9}$	$27.1^{+3.5}_{-3.1}$
				67	$(-5.00°, 5.00°), -3.25°$	$1.88^{+0.43}_{-0.35}$	$50.6^{+7.0}_{-6.3}$	$18.8^{+2.6}_{-2.3}$
				58	$(-5.00°, 5.00°), -3.75°$	$1.52^{+0.26}_{-0.23}$	$40.3^{+5.8}_{-5.2}$	$15.7^{+2.2}_{-2.0}$

Table 5.1 (continued)

Project	Field deg²	Method	$t_{E,max}$ days	N_{event} events	l, b^*	τ ×10⁶	Γ_{deg^2} yr⁻¹deg⁻²	Γ_{star} ×10⁻⁶yr⁻¹star⁻¹
				43	(−5.00°,5.00°), −4.25°	$1.47^{+0.32}_{-0.26}$	$28.6^{+4.9}_{-4.2}$	$11.6^{+2.0}_{-1.7}$
				22	(−5.00°,5.00°), −4.74°	$0.76^{+0.22}_{-0.18}$	$15.0^{+3.6}_{-2.9}$	$6.6^{+1.6}_{-1.3}$
				16	(−5.00°,5.00°), −5.23°	$0.94^{+0.39}_{-0.28}$	$16.2^{+4.6}_{-3.6}$	$7.6^{+2.2}_{-1.7}$
				8	(−5.00°,5.00°), −5.72°	$1.34^{+0.84}_{-0.51}$	$13.4^{+5.8}_{-4.0}$	$7.2^{+3.1}_{-2.1}$
				4	(−5.00°,5.00°), −6.23°	$0.85^{+0.64}_{-0.35}$	$23.7^{+19.5}_{-10.4}$	$13.9^{+11.5}_{-6.1}$
		RCG	200	16	(−5.00°,5.00°), −1.69°	$2.87^{+1.03}_{-0.75}$	$9.8^{+2.8}_{-2.2}$	$47.3^{+13.6}_{-10.6}$
				16	(−5.00°,5.00°), −2.26°	$3.44^{+1.45}_{-1.03}$	$7.5^{+2.2}_{-1.7}$	$38.7^{+11.2}_{-8.7}$
				11	(−5.00°,5.00°), −2.76°	$1.40^{+0.67}_{-0.45}$	$3.9^{+1.4}_{-1.0}$	$20.9^{+7.4}_{-5.5}$
				14	(−5.00°,5.00°), −3.25°	$1.93^{+0.76}_{-0.55}$	$4.3^{+1.3}_{-1.0}$	$23.6^{+7.3}_{-5.6}$
				11	(−5.00°,5.00°), −3.75°	$1.95^{+0.98}_{-0.64}$	$3.5^{+1.3}_{-0.9}$	$21.2^{+7.7}_{-5.7}$
				4	(−5.00°,5.00°), −4.25°	$0.57^{+0.55}_{-0.26}$	$1.1^{+0.8}_{-0.4}$	$7.3^{+5.0}_{-2.9}$
				3	(−5.00°,5.00°), −5.15°	$0.93^{+0.99}_{-0.41}$	$0.6^{+0.5}_{-0.3}$	$4.7^{+3.9}_{-2.4}$

*The values of Galactic latitude (l) and Galactic longitude (b) shown in Table 5.1 are average position of the map or average Galactic latitude of field in each Galactic longitude bin.

Note [1] Udalski et al. (1994), [2] Alcock et al. (1997), [3] Alcock et al. (2000), [4] Afonso et al. (2003), [5] Sumi et al. 2003, [6] Popowski et al. 2005, [7] Sumi et al. (2006), [8] (Hamadache et al. (2006), [9] Sumi et al. (2013)

four main populations are assumed: a thin disk; thick disk; bulge/bar; and stellar halo. The modelling of each population is based on theoretical consideration (stellar formation, evolution and density distributions) and is constrained by empirical observations (local luminosity function, star counts and proper-motion surveys). A star age, IMF and SFR are assumed, allowing the model to generate the distribution function of absolute magnitude, age and effective temperature of the stars. In order to obtain a more realistic model, the observational errors and Poisson noise are included.

The model also includes a 3D extinction map (Marshall et al. 2006). An interstellar extinction distribution in three dimensions is computed from the 2MASS survey (Cutri et al. 2003) towards the inner Galaxy ($|l| \leq 100°$ and $|b| \leq 10°$), with $15'$ resolution. Marshall et al. (2006) calculated the extinction as a function of distance along each line of sight by comparing observed reddened stars to unreddened simulated stars from the Besançon model. This distribution can be used to determine the observed colours and magnitudes of the simulated stars. In the following work, a later version of Besançon model (Robin et al. 2014) has been used and we summarize the main components below.

5.4.1 Thin Disk

The thin disk is a major component in the Galactic central region. It is assumed to have an age of 10 Gyr. A constant SFR over the past 10 Gyr is assumed, along with an IMF with two slopes, $dN/dm \propto m^{-1.6}$ for $m < 1M_\odot$ and $dN/dm \propto m^{-3.0}$ for $m > 1M_\odot$. The total mass of the thin disk is $9.3 \times 10^9 \, M_\odot$. The luminosity function determined from *Hipparcos* observations is adopted (Haywood et al. 1997a, b; Robin et al. 2003), whilst the underlying density law follows the Einasto (1979) density profile with a central hole:

$$\rho_d = \rho_{d_0} \times \left[\exp\left(-\sqrt{0.25 + \left(\frac{R}{R_d}\right)^2 + \left(\frac{Z}{\epsilon R_d}\right)^2}\right) \right.$$
$$\left. -\exp\left(-\sqrt{0.25 + \left(\frac{R}{R_h}\right)^2 + \left(\frac{Z}{\epsilon R_h}\right)^2}\right) \right], \qquad (5.17)$$

where

- R and Z are the cylindrical galactocentric coordinates;
- ϵ is the axis ratio of the ellipsoid
- R_d is the scale length of the disk
- R_h is the scale length of the central hole in the disk
- ρ_{d_0} is the mass density normalization, assuming that the Sun is located at $R_\odot = 8$ kpc from the Galactic centre and $Z_\odot = 15$ pc above the disk plane (Jahreiß and Wielen 1997)

The important parameters of the thin disk are the scale height and the size of the central hole. The disk scale height is constrained by the Galactic potential, stellar evolutionary tracks and star counts which are used to constrain the initial mass function (IMF) and star formation rate (SFR) of the disk population (Haywood et al. 1997b). Due to a hole in the centre. the maximum density of the thin disk is located at about 2.5 kpc from the Galactic centre. The kinematics follows the *Hipparcos* empirical estimates of Gomez et al. (1997). The population of thin disk is divided into 7 distinct components with different distributions of age, scale height and velocities (Robin et al. 2012).

5.4.2 Thick Disk

The thick disk is of much lower density than the thin disk locally but becomes important at Galactic latitudes above about 8–10°. In the model it is assumed to be a separate population from the thin disk, with a distinct star formation history. Recent constraints from SDSS and 2MASS data lead to revisions of the scale length and scale heights (Robin et al. 2014). We make use of the single thick disk episode of formation presented in Robin et al. (2014), modelled by a 12 Gyr isochrone of metallicity -0.78 dex, with a density law following a modified exponential (parabola up to z = 658 pc, followed by an exponential with a scale height of 533 pc), which is roughly equivalent to a $sech^2$ function of scale height 450 pc. Its properties have been constrained by the star count results of Reylé and Robin (2001) with an IMF of $dN/dm \propto m^{-0.5}$. The radial density follows an exponential with a scale length of 2.355 kpc. Its kinematics follow the results of Ojha et al. (1996).

5.4.3 Bulge/Bar

A new model of the bulge of the Besançon model has been proposed by Robin et al. (2012), as the sum of two ellipsoids: a standard boxy bulge (bar), the most massive component which dominates the stellar content of latitudes below about 5°, and another ellipsoid (thick bulge) with longer and thicker structure which can be observed at higher latitudes where the bar starts to be less prominent. However, Robin et al. (2014) showed that the "thick bulge" population was in fact the inner part of the thick disk, where its short scale length makes a large contribution in the bulge region. Hence, in this new version, the populations in the bulge region are: the thin disk, the bar and the thick disk. The angle of the bar to the Sun-Galactic Centre direction is 13°. The bar kinematics are taken from the model of Fux (1999) and the bulge kinematics are established to reproduce the BRAVA survey data (Rich et al. 2007). The stellar density and luminosity function are taken from the results of Picaud and Robin (2004) with a single burst population of 10 Gyr age. The IMF below and above 0.7 $M_\odot$ are assumed to be $dN/dm \propto m^{-1.5}$ and a Salpeter slope, $dN/dm \propto m^{-2.35}$, respectively

(Picaud and Robin 2004). The total bar mass is $5.9 \times 10^9 \, M_\odot$. The model mass to light ratio is 2.0 at the Sagittarius Window Eclipsing Exoplanet Search (SWEEPS) field ($l = 1.25°$, $b = -2.65°$) in Johnson-I band which is compatible with result of Calamida et al. (2015) in F814W filter (wide I).

5.4.4 Stellar Halo

The stellar halo is older than the thick disk (14 Gyr) and metal poor ([Fe/H] = -1.78). A single burst population with an IMF, $dN/dm \propto m^{-0.5}$, and total mass of $4.0 \times 10^{10} \, M_\odot$ are assumed (Robin et al. 2003). The density law has been revised in the study of SDSS+2MASS star counts (Robin et al. 2014). It is now modelled with a power law density with an exponent of 3.39 and an axis ratio of 0.768. Its kinematics is modelled with Gaussian velocities distributions of dispersion $(131, 106, 85)$ in km/s in the (U,V,W) plane, and no rotation.

5.5 Manchester–Besançon Microlensing Simulator

The Manchester–Besançon Microlensing Simulator (MaBμlS) is the first real-time web-based program that can create microlensing maps of the Galactic bulge.[2] The input parameters of the program are summarized in Table 5.2.

From the MaBμlS web page, the parameters can be selected as in Fig. 5.3. After selecting the parameters, the program uses pre-computed simulated data to compute microlensing property maps. In the program, microlensing maps of the Galactic bulge in an area between -10 and 10 degrees from the Galactic centre for both Galactic latitude and longitude with 15' × 15' resolution are simulated. In order to obtain enough sample stars in all magnitude ranges for simulating microlensing maps of the Galactic bulge, the simulated stars from the Besançon model are divided into 4 magnitude ranges catalogues : $-10 \le H < 15$, $15 \le H < 19$, $19 \le H < 23$ and $H > 23$. Due to computational time limits, each catalogue has a different solid angle size. The sizes of solid angle are calculated to contain around 1,000 stars at Baade's Window ($l = 1°$, $b = 4°$) (Table 5.3).

All source and lens pairs within the same field are used to calculate the maps of optical depth, average Einstein crossing time and microlensing event rate, for both the resolved source and difference imaging analysis (DIA) source (Alard 2000; Wozniak 2000; Bramich 2008) method. For each line of sight, the computational time for each microlensing property is longer than 1 minute. Therefore, it is impossible to create an interactive web-based microlensing map, which contains a maximum 6,400 line of sights, by real-time computation. In order to create the real-time web-based

---

[2]Manchester–Besançon Microlensing Simulator - MaBμlS, which is publicly available online at http://www.mabuls.net/.

Table 5.2 Input parameters of the MaBμlS

Property	Value
Version of Besançon model	Version 1307
Galactic latitude (b)	$-10.125° - +9.875°$
Galactic longitude (l)	$-9.875° - +10.125°$
Filter	$U, V, B, R, I, J, H, K, L$
Microlensing source	Resolved sources, DIA sources
Minimum source apparent magnitude ($M_{\min}$)	≥ 12 or no limit
Maximum source apparent magnitude ($M_{\max}$)	≤ 23
Event duration (t_E)	1–1,000 day
Lens population	All, Thin disk and thick disk, Bulge and halo
Microlensing property	Optical depth, Average Einstein radius crossing time
	Event rate per sky area, Event rate per source star
Microlensing map image parameter	Interpolation
	Number of contours
	Power of colour bar stretch power law
	Percentile of intensity clipping
Error map image parameter	Interpolation
	Number of contours
	Power of colour bar stretch power law
	Percentile of intensity clipping

Table 5.3 Solid angles of the MaBμlS simulated catalogues

Magnitude range	Solid angel (deg^2)
$-10 \leq H < 15$	4.5×10^{-3}
$15 \leq H < 19$	1.4×10^{-4}
$19 \leq H < 23$	1.6×10^{-5}
$H > 23$	2.0×10^{-5}

calculation, the pre-computed microlensing properties are stored in 4 dimensioned grids:

- **Filter**: 9 filters: $U, V, B, R, I, J, H, K, L$
- **Magnitude limit**: 12 ranges: $M_s <$ 12.0, 13.0, 14.0, 15.0, 16.0, 17.0, 18.0, 19.0, 20.0, 21.0, 22.0, 23.0
- **Crossing time**: 9 ranges: $(t_{E,min}, t_{E,max}) = (1, 10^{1/3})$ d, $(10^{1/3}, 10^{2/3})$ d, $(10^{2/3}, 10)$ d, $(10, 10^{4/3})$ d, $(10^{4/3}, 10^{5/3})$ d, $(10^{5/3}, 100)$ d, $(100, 10^{7/3})$ d, $(10^{7/3}, 10^{8/3})$ d, $(10^{8/3}, 1000)$ d
- **Population**: 2 populations: Thin disk and thick disk lenses, Bulge and Halo lenses

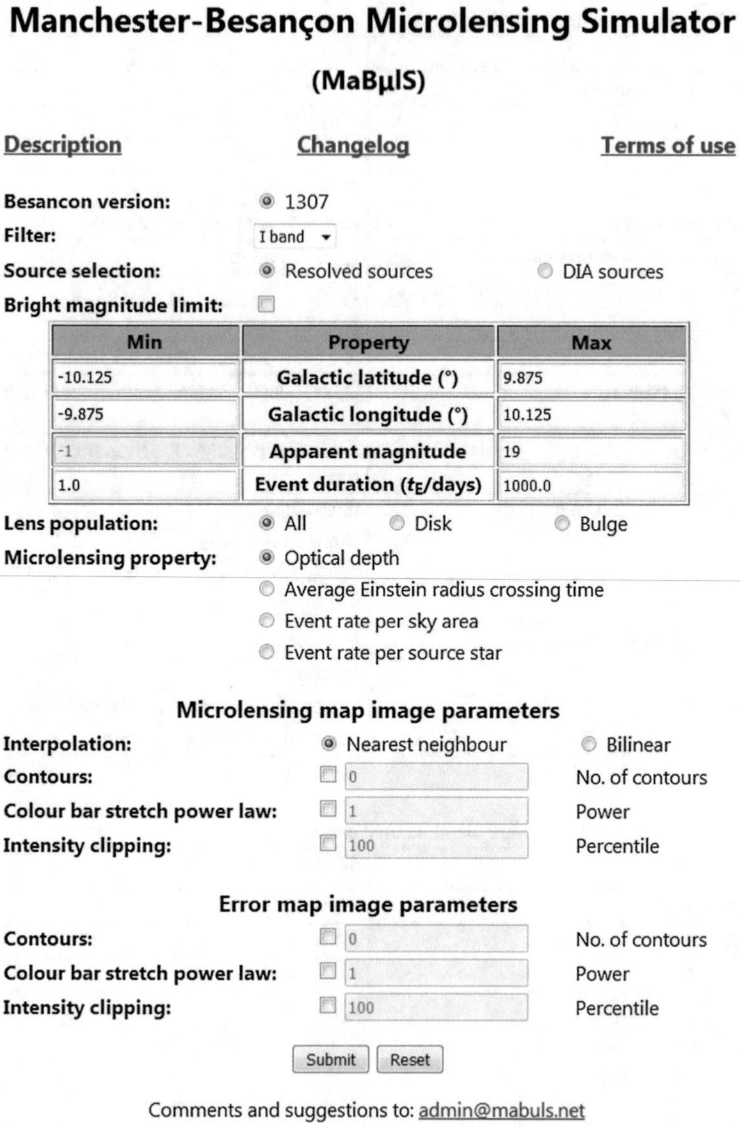

Manchester-Besançon Microlensing Simulator

(MaBμlS)

Besancon version: ◉ 1307
Filter: I band ▾
Source selection: ◉ Resolved sources   ○ DIA sources
Bright magnitude limit: ☐

Min	Property	Max
-10.125	**Galactic latitude (°)**	9.875
-9.875	**Galactic longitude (°)**	10.125
-1	**Apparent magnitude**	19
1.0	**Event duration (t_E/days)**	1000.0

Lens population: ◉ All   ○ Disk   ○ Bulge
Microlensing property: ◉ Optical depth
                               ○ Average Einstein radius crossing time
                               ○ Event rate per sky area
                               ○ Event rate per source star

Microlensing map image parameters

Interpolation: ◉ Nearest neighbour   ○ Bilinear
Contours: ☐ 0   No. of contours
Colour bar stretch power law: ☐ 1   Power
Intensity clipping: ☐ 100   Percentile

Error map image parameters

Contours: ☐ 0   No. of contours
Colour bar stretch power law: ☐ 1   Power
Intensity clipping: ☐ 100   Percentile

Submit   Reset

Comments and suggestions to: admin@mabuls.net

Fig. 5.3 The MaBμlS parameter selections web-page

For unresolved sources, the instantaneous fraction of events with impact parameter u small enough to be detectable scales as u^2, though over time the rate of detectable events scales as u. For calculating microlensing parameters for DIA sources, the optical depth and average time scale are weighted by $u^2_{\tau,min}=\min(1, u^2)$ and $u_{t,min}=\min(1, u)$, respectively, where u is

$$u = \sqrt{\frac{2A}{\sqrt{A^2 - 1}} - 2} \,, \tag{5.18}$$

where A is minimum magnification needed for detection,

$$A = 10^{0.4(M - M_{lim})} \,. \tag{5.19}$$

In Eq. 5.19, M is the survey magnitude of the source star and M_{lim} is the magnitude limit in the same filter. Therefore, weighted minimum impact parameters for DIA sources can written as:

$$u_{min} = \begin{cases} 1, & A \le (3/\sqrt{5}) \,, \\ \sqrt{\frac{2A}{\sqrt{A^2-1}} - 2}, & \text{otherwise} \,. \end{cases} \tag{5.20}$$

For resolved sources, the minimum impact parameter can be written as

$$u_{min} = \begin{cases} 0, & M_s > M_{lim} \,, \\ 1, & \text{otherwise} \,, \end{cases} \tag{5.21}$$

for both optical depth and average time scale.

In following work, the finite source effect has been taken into account. Only events in which the source star angular radius is smaller than the minimum impact distance are used. Therefore, the weight can be written as

$$u_{min} = \begin{cases} 0, & \theta_E \times u_{min} \le R_s/D_s \,, \\ u_{min}, & \text{otherwise} \,, \end{cases} \tag{5.22}$$

where R_s is stellar radius of source star.

The average impact parameter is weighted by $\mu D_l R_E$ because of the different frequency of occurrence of microlensing events with different relative proper motion (μ), lens distance (D_l) and Einstein radius (R_E) in each pair. The weighted average can be written as,

$$\langle u_{min} \rangle_w = \begin{cases} 1, & M_s < M_{lim} \,, \\ \dfrac{\sum_{l=1}^{N_l(D_s>D_l)} u_{min}\mu D_l R_E}{\sum_{l=1}^{N_l(D_s>D_l)} \mu D_l R_E}, & \text{otherwise} \,. \end{cases} \tag{5.23}$$

The weighted average is used to weight the parameters as the average impact parameter of all lenses for each source. For the real-time computation, pre-computed results are stored in from of a ratio of two function $S(x)$ and $W(x)$ with the final result being $S(x)/W(x)$. For the pre-computed optical depth we define,

$$S(\tau) = \sum_{s=1}^{N_s} \sum_{l=1}^{N_l(D_s > D_l)} u_{\tau,min}^2 \pi \theta_E^2 \frac{\Omega_0}{\Omega_s \Omega_l} , \qquad (5.24)$$

and

$$W(\tau) = \sum_{s=1}^{N_s} \left\langle u_{\tau,min}^2 \right\rangle_w \frac{\Omega_0}{\Omega_s} , \qquad (5.25)$$

where Ω_0 is solid angle of the field. Ω_l and Ω_s are the solid angles over which the source and lens catalogues are simulated, respectively. Hence $\tau = S(\tau)/W(\tau)$, For the average crossing time, we have

$$S(t_E) = \sum_{s=1}^{N_s} \sum_{l=1}^{N_l(D_s > D_l)} u_{t,min} D_l^2 \theta_E^2 \frac{\Omega_0}{\Omega_s \Omega_l} , \qquad (5.26)$$

and

$$W(t_E) = \sum_{s=1}^{N_s} \sum_{l=1}^{N_l(D_s > D_l)} \left\langle u_{t,min} \right\rangle_w D_l^2 \theta_E \mu \frac{\Omega_0}{\Omega_s \Omega_l} . \qquad (5.27)$$

In MaBμlS, the specific range of sources magnitude and Einstein crossing time can be selected. For a specific range, the event rate is calculated by the ratio for the summed rate weights $\left\langle u_{t,min} \right\rangle_w D_l^2 \theta_E \mu$ of all events and the sum only in the selected range. The calculated equation for the selected range can be written as:

- Optical depth

$$\tau = \frac{\sum_i^{\text{All bins}} S_i(\tau) \sum_j^{\text{Selected bins}} S_j(t_E)}{\sum_i^{\text{All bins}} W_i(\tau) \sum_i^{\text{All bins}} S_i(t_E)} . \qquad (5.28)$$

- Average Einstein radius crossing time

$$\langle t_E \rangle = \frac{\sum_i^{\text{Selected bins}} S_i(t_E)}{\sum_i^{\text{Selected bins}} W_i(t_E)} . \qquad (5.29)$$

- Event rate per sky area

$$\Gamma_{\deg^2} = \frac{2}{\pi \Omega_0} \frac{\sum_i^{\text{All bins}} S_i(\tau) \sum_j^{\text{Select bins}} W_j(t_E)}{\sum_i^{\text{All bins}} S_i(t_E)} . \qquad (5.30)$$

- Event rate per source star

$$\Gamma_{\text{star}} = \frac{2}{\pi} \frac{\sum_i^{\text{All bins}} S_i(\tau) \sum_j^{\text{Select bins}} W_j(t_E)}{\sum_i^{\text{All bins}} S_i(t_E) \sum_i^{\text{All bins}} W_i(\tau)} . \qquad (5.31)$$

where bin is the pre-computed grid. However, the data were computed at specific grid points of location, source magnitude and crossing time. To obtain parameters of a point in between the grids, linear interpolation between two nearest grids is performed.

The error maps of the properties are useful for estimating the accuracy of the property maps in a specific filter, source magnitude and crossing time range. The estimated errors of each quantity are calculated by separating the quantities into two groups by random number and computing their standard deviation. The uniform distribution between 0 and 1 is used to be the random number for each star. If the random number of a star less than 0.5, it belongs to group 1. The stars with random number greater than or equal to 0.5 belong to group 2. In each group, the parameters are calculated with the same methods as the properties maps. The standard deviation of parameter from each group is used to be the error of the parameter. These estimated errors show the variability of quantities from the Besançon Galactic model.

For the computation, some lines of sight contain a large number of stars, especially in low Galactic latitude areas which consume large amounts of computational time. The calculation with 1,000 stars from each catalogue provides a good sample of the stars in that area. The 1,000 star sample has a normalised estimated error of less than 30% for an $H < 20$ source magnitude. Therefore, in this work, for the lines of sight that have catalogues exceeding 1,000 stars, only 1,000 randomly sampled stars in each catalogue are used. The solid angles are correspondingly renormalized.

The computed microlensing maps are shown from http://www.mabuls.net as in Fig. 5.4 Several examples of microlensing map outputs are shown in Figs. 5.5, 5.6, 5.7, 5.8 and 5.9.

For each map, the user can investigate the microlensing rate as a function of source magnitude and Einstein crossing time in 1×1 degree regions as in Fig. 5.10. A computed event rate histogram for such a region is shown as in Fig. 5.11. The example histogram of DIA sources with $H < 20$ as function of source magnitude and crossing time are shown in Figs. 5.12 and 5.13, respectively.

Fig. 5.4 Example MaBμlS
microlensing map output
web-page

Manchester-Besançon Microlensing Simulator
(MaBμlS)

Description   Changelog   Terms of use

Input parameters		
Microlensing property	Optical depth	
Method	Resolved sources	
Filter	I band	
Galactic latitude	-9.875	-	10.125
Galactic longitude	-10.125	-	9.875
Magnitude	<	19
Einstien crossing time (days)	1.0	-	1000.0
Lens population	All lenses	

Microlensing map

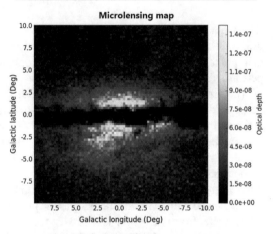

You can save the image above by right-clicking on it in your browser. Or, you can
download this map as a data file to make your own plot.

Error map

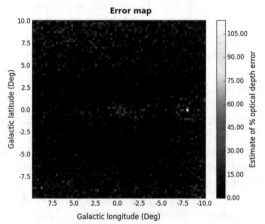

Time-scale distribution

Investigate

Comments and suggestions to: admin@mabuls.net

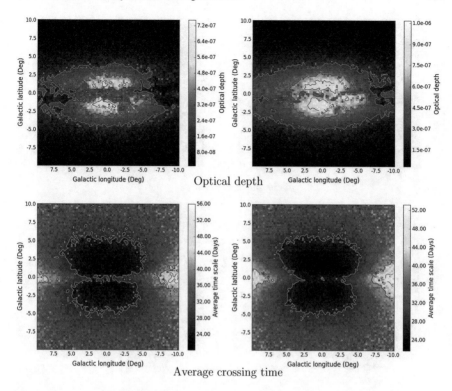

Fig. 5.5 Example optical depth, average crossing time and event rate maps with resolved sources (Left) and DIA sources (Right) brighter than $H < 20$ from MaBμlS

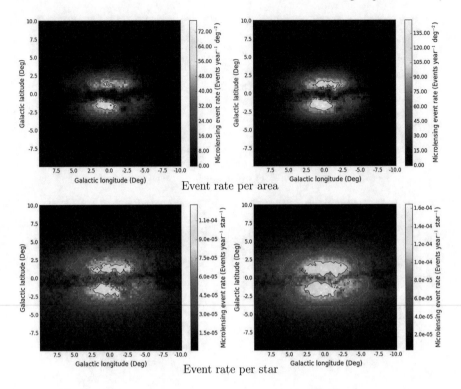

Event rate per area

Event rate per star

Fig. 5.5 (continued)

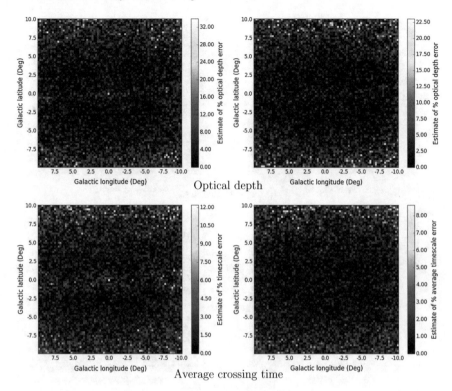

Fig. 5.6 Example optical depth, average crossing time and event rate variance maps with resolved sources (Left) and DIA sources (Right) brighter than $H < 20$ from MaBμlS

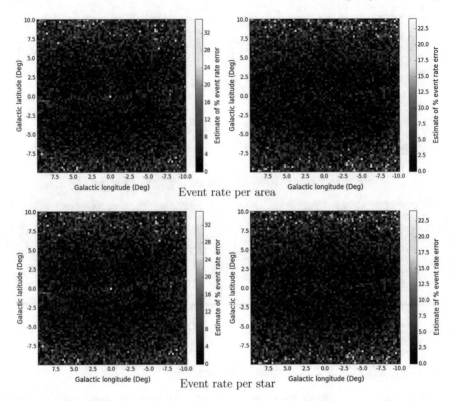

Fig. 5.6   (continued)

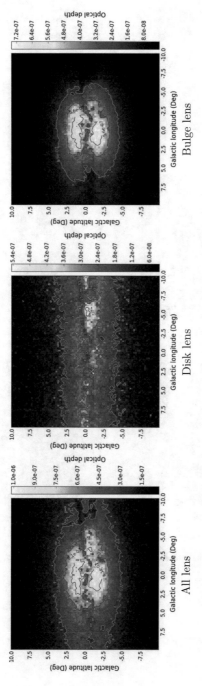

Fig. 5.7 Example optical depth of all, disk and bulge lens with DIA sources brighter than $H < 20$ from MaBμlS

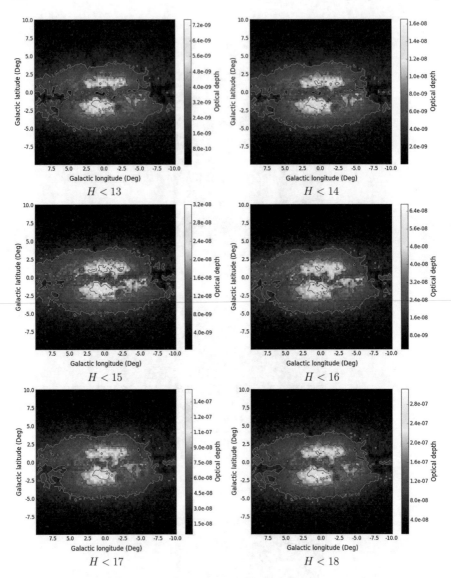

Fig. 5.8 Example optical depth of DIA sources at various magnitude limit from MaBµlS

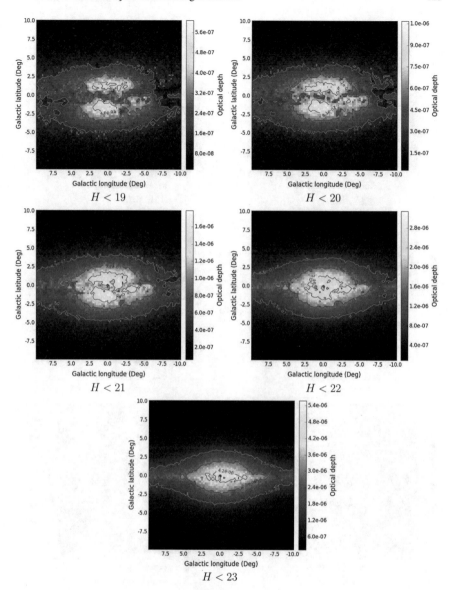

Fig. 5.8 (continued)

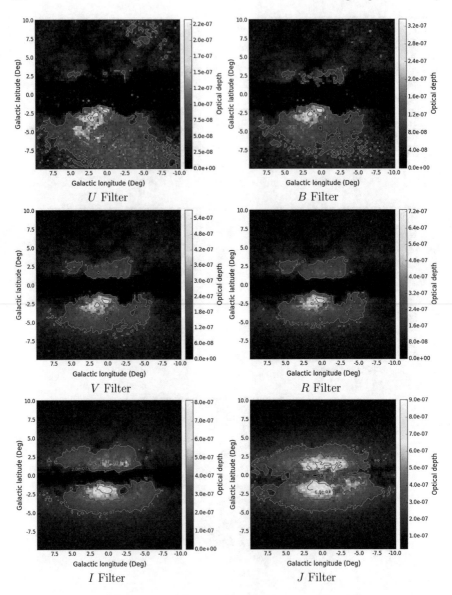

Fig. 5.9 Example optical depth of DIA sources for various filters at $< 20^{th}$ magnitude from MaBμlS

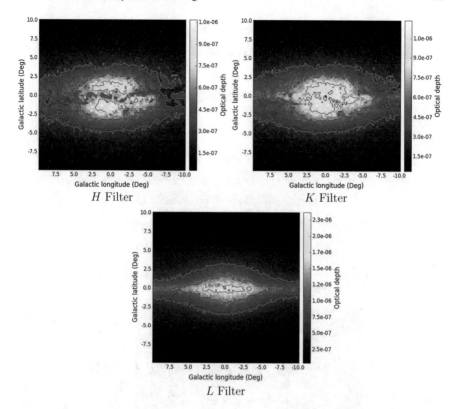

H Filter

K Filter

L Filter

Fig. 5.9   (continued)

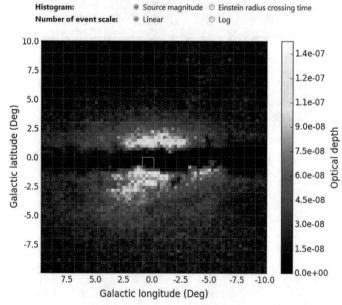

Fig. 5.10 Example MaBμlS timescale and magnitude histogram selection web-page (beta version)

Manchester-Besançon Microlensing Simulator

(MaBμlS)

| Description | Changelog | Terms of use |

Input parameters		
Method	Resolved sources	
Filter	I band	
Galactic latitude	-0.875	-	0.125
Galactic longitude	-0.125	-	0.875
Einstien crossing time (days)	1.0	-	1000.0
Lens population	All lenses	

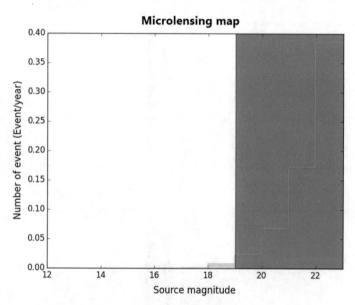

You can save the image above by right-clicking on it in your browser. Or, you can download this map as a data file to make your own plot.

Comments and suggestions to: admin@mabuls.net

Fig. 5.11 Example MaBμlS event rate histogram result web-page

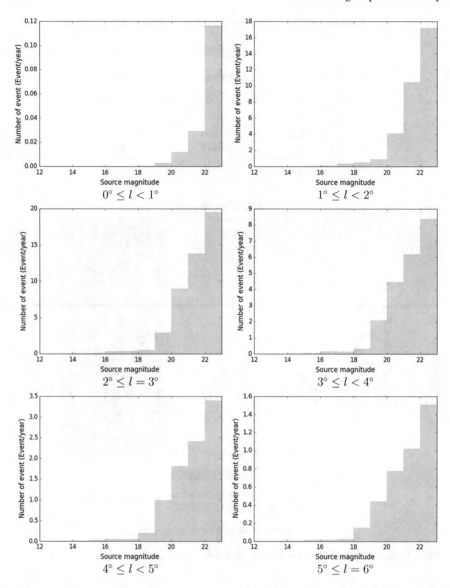

Fig. 5.12 Histograms of microlensing event rate as a function of source magnitude in H-band at $0° \leq b < 1°$ with different Galactic longitude ranges from MaBμlS

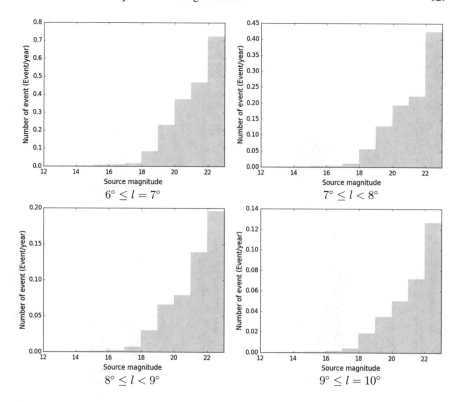

Fig. 5.12 (continued)

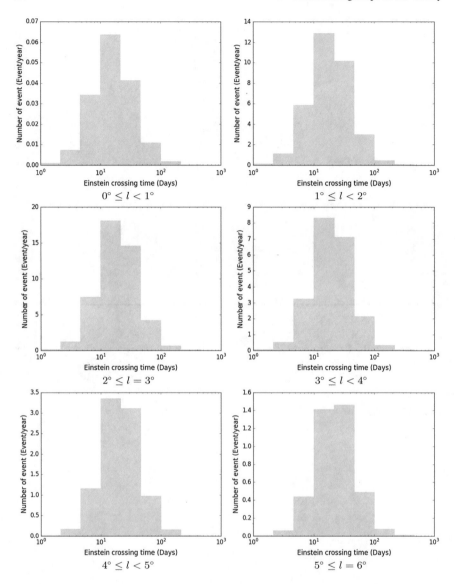

Fig. 5.13 Histograms of microlensing event rate as a function of crossing time in H-band at $0° \leq b < 1°$ with different Galactic longitude ranges from MaBμlS

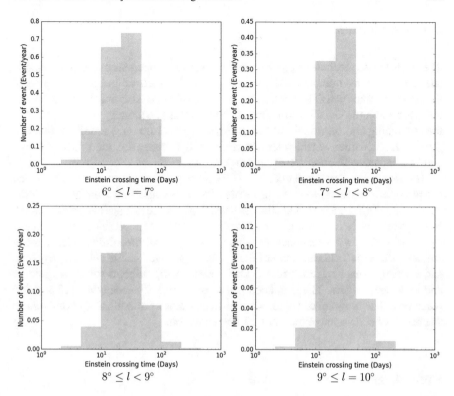

Fig. 5.13 (continued)

5.6   Conclusion

The microlensing technique can be used to discover exoplanets and study Galactic structure. To date, microlensing surveys toward the Galactic bulge have detected thousands of microlensing events. The statistics of detected events provides useful information on the Galactic structure, such as its mass and kinematics. As there are many ongoing and future microlensing surveys, a microlensing simulator, that can provide detailed microlensing maps, is useful for optimising the future survey design and to compare with observed event sample.

   The Besançon Galactic model is a Galactic population synthesis model, which can provide the observable stellar properties, based on constraints from observational data. Using the Besançon model, the microlensing properties maps can be calculated. In this work, we have developed the Manchester–Besançon microlensing simulator (MaBμlS - http://www.mabuls.net), which is the first real-time online microlensing simulator. The MaBμlS can simulate the optical depth, average Einstein crossing time and event rate maps of resolved and DIA sources with different source magnitude and in different filters. The simulated maps cover $10 \times 10°$ area with $0.25 \times 0.25°$ resolution. The program can also calculate histograms of microlensing event rate as a function of source magnitude in 1×1 degree region.

References

Afonso C, Albert JN, Alard C et al (2003) A & A 404:145
Alard C (2000) A & AS 144:363
Alcock C, Allsman RA, Alves D et al (1997) ApJ 479:119
Alcock C, Allsman RA, Alves DR et al (2000) ApJ 541:734
Awiphan S, Kerins E, Robin AC (2016b) MNRAS 456:1666
Bond IA, Rattenbury NJ, Skuljan J et al (2002) MNRAS 333:71
Bond IA, Udalski A, Jaroszyński M et al (2004) ApJ 606:L155
Bramich DM (2008) MNRAS 386:L77
Calamida A, Sahu KC, Casertano S et al (2015) ApJ 810:8
Cutri RM, Skrutskie MF, van Dyk S et al. (2003) 2MASS All sky catalog of point sources
Einasto J (1979) In: Burton WB (ed) The large-scale characteristics of the galaxy, vol 84 of IAU Symposium, pp. 451–458
Einstein A (1915) Sitzungsber. preuss. Akad. Wiss. 47(2): 831–839
Einstein A (1936) Science 84:506
Fux R (1999) A & A 345:787
Gaudi BS (2012) ARA & A 50:411
Gomez AE, Grenier S, Udry S et al. (1997). In: Bonnet RM, Høg E, Bernacca PL, Emiliani L, Blaauw A, Turon C, Kovalevsky J, Lindegren L, Hassan H, Bouffard M, Strim B, Heger D, Perryman MAC, Woltjer L (eds) Hipparcos - Venice '97, vol 402 of ESA Special Publication, pp. 621–624
Hamadache C, Le Guillou L, Tisserand P et al (2006) A & A 454:185
Han C, Gould A (2003) ApJ 592:172
Haywood M, Robin AC, Creze M (1997a) A & A 320:428
Haywood M, Robin AC, Creze M (1997b) A & A 320:440

Henderson CB, Gaudi BS, Han C et al (2014) ApJ 794:52

Hwang K-H, Han C, Choi J-Y et al. (2015). arXiv:1507.05361

Jahreiß H, Wielen R (1997). In: Bonnet RM, Høg E, Bernacca PL, Emiliani L, Blaauw A, Turon C, Kovalevsky J, Lindegren L, Hassan H, Bouffard M, Strim B, Heger D, Perryman MAC, Woltjer L (eds) Hipparcos - Venice '97, vol. 402 of ESA special publication, pp. 675–680

Kennedy GM, Kenyon SJ (2008) ApJ 682:1264

Kerins E, Robin AC, Marshall DJ (2009) MNRAS 396:1202

Laureijs R, Amiaux J, Arduini S et al. (2011), Euclid Definition Study Report, ESA

Lecar M, Podolak M, Sasselov D, Chiang E (2006) ApJ 640:1115

Liebes S (1964) Phys Rev 133:835

Marshall DJ, Robin AC, Reylé C, Schultheis M, Picaud S (2006) A & A 453:635

Mróz P, Udalski A, Skowron J et al (2017) Nature 548:183

Ojha DK, Bienayme O, Robin AC, Creze M, Mohan V (1996) A & A 311:456

Paczynski B (1996) ARA & A 34:419

Picaud S, Robin AC (2004) A & A 428:891

Popowski P, Griest K, Thomas CL et al (2005) ApJ 631:879

Refsdal S (1964) MNRAS 128:295

Reylé C, Robin AC (2001) A & A 373:886

Rich RM, Reitzel DB, Howard CD, Zhao H (2007) ApJ 658:L29

Robin AC, Reylé C, Derrière S, Picaud S (2003) A & A 409:523

Robin AC, Marshall DJ, Schultheis M, Reylé C (2012) A & A 538:A106

Robin AC, Reylé C, Fliri J, Czekaj M, Robert CP, Martins AMM (2014) A & A 569:A13

Spergel D, Gehrels N, Baltay C et al (2015). arXiv:1503.03757

Spergel D, Gehrels N, Breckinridge J et al (2013). arXiv:1305.5422

Sumi T, Abe F, Bond IA et al (2003) ApJ 591:204

Sumi T, Woźniak PR, Udalski A et al (2006) ApJ 636:240

Sumi T, Bennett DP, Bond IA et al (2013) ApJ 778:150

Udalski A, Szymanski M, Mao S et al (1994) ApJ 436:L103

Udalski A, Szymanski MK, Soszynski I, Poleski R (2008) Acta Astronaut 58:69

Wood A, Mao S (2005) MNRAS 362:945

Wozniak PR (2000) Acta Astronaut 50:421

Wyrzykowski Ł, Rynkiewicz AE, Skowron J et al (2015) ApJS 216:12

Chapter 6
Besançon Model Simulations of MOA-II

To date, there are thousands of microlensing events have been detected by survey teams, such as MOA (Bond et al. 2002), OGLE (Udalski et al. 2008) and KMTNet (Hwang et al. 2015). These event samples can be used to calculate the optical depth, average crossing time and event rate of the microlensing events at the Galactic bulge. In a recent study, the MOA-II survey (Sumi et al. 2013) determined these parameters from a study of 474 microlensing events in 2006–2007 observation season.

In this chapter, we present the first field-by-field comparison between microlensing observations, MOA-II (Sumi et al. 2013), and the Besançon population synthesis Galactic model. Using a recent version of Besançon Galactic model (Robin et al. 2014) and the Manchester-Besançon Microlensing Simulator (MaBIS) (See Chap. 5), maps of optical depth, average event duration and event rate for resolved source populations and for difference imaging analysis (DIA) events are provided. The simulation follows the selection criteria of the MOA-II survey. The predicted event time-scale distributions are also compared to that observed.

6.1 MOA-II Microlensing Survey

The MOA survey started observations with a 0.61 metre telescope at Mt. John observatory, New Zealand, in 1995 (Sumi 2010). The MOA survey is designed to perform a long-term continuous wide-field survey at the Galactic bulge in order to find and alert microlensing events. The second phase of the survey (MOA-II) uses the 1.8 metre telescope at Mt. John observatory. The telescope is equipped with MOA-cam3 with ten $2K \times 4K$ CCDs and $15 \mu m$ pixels (Sako et al. 2008). The plate scale is 0.58 arcsec per pixel which gives $2.18 \deg^2$ field-of-view. The survey covers $4 \deg^2$ every 10 min. The custom MOA-Red wide-band filter, the sum of the flux measured in standard Kron/Cousins R and I band, are used. The data are reduced by the DIA method (Bond et al. 2001).

© Springer International Publishing AG, part of Springer Nature 2018
S. Awiphan, *Exomoons to Galactic Structure*, Springer Theses,
https://doi.org/10.1007/978-3-319-90957-8_6

Sumi et al. (2013) published microlensing parameter maps of the MOA-II 2006–2007 observational season. The survey delivered a high cadence photometry of millions of stars in 21 fields towards the Galactic bulge. Each field is divided into 80 sub-fields, however one sub-field is not used due to technical reasons. The number of used sub-fields is 1536 in total. The data in fields gb5 and gb9 were taken with 10 min cadence which provided 8,250 images in each field. For the 19 other fields, 1660–2980 images of each were taken with 50 min cadence.

In order to perform microlensing event selection, light curves with a single instantaneously bright event and a flat baseline, which are fit with single lens microlensing model (See Sect. 5.1.2), are chosen (Sumi et al. 2011). Events should have a minimum impact parameter $u_0 < 1.0$. From previous studies, binary lenses contribute about 8% (Jaroszynski 2002), 6% (Alcock et al. 2000), 3% (Jaroszynski et al. 2004) and 6% (Sumi et al. 2006) of the event rate. Sumi et al. (2013) excluded binary lens events and applied a 6% correction. From the data, more than 1,000 microlensing candidates are detected, but only 474 high quality events pass the criteria (Fig. 6.1, Sumi et al. 2011). Three event selection criteria are adopted:

- **Cut-0**: Difference image event peaks have signal to noise ratio > 5 at the same positions on more than 2 images.
- **Cut-1**: The objects have more than 500 data points with more than 10 baseline data points (χ^2/dof $\leq$ 3). The peak contains more than 3 consecutive data points with excess flux > $3\sigma_i \sqrt{\chi^2/\text{dof}}$, where σ_i is the error of the flux for the ith measurement.

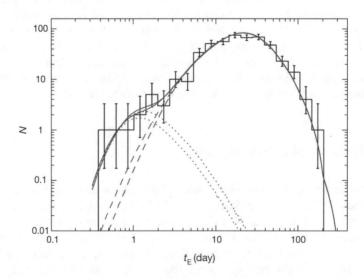

Fig. 6.1 Histogram of MOA-II 474 high quality events with the best-fit models with the power-law (red) and log-normal (blue) mass functions for stellar, stellar remnant and brown dwarf populations (dashed lines), planetary-mass population (dotted lines) and the sums of these populations (solid lines) (Sumi et al. 2011)

- **Cut-2**: The events can be modelled with a point source and a single lens object with $\chi^2/\text{dof} \leq 2$.

Sumi et al. (2013) defined 2 source samples; an all-source sample and a Red Clump Giant (RCG) sample. The all-source sample is all 474 events which have peak magnitude $I < 20$. The RCG sample is 83 events which have baseline magnitude $I = 17.5$ mag and $(V - I) > (V - I)_{RC} - 0.3$ mag, where $(V - I)_{RC}$ is the $V - I$ colour of the RCG centroid which vary with on location of the sub-field. The RCG controid colours are calculated by the method of Nataf et al. (2013).

In order to compute microlensing properties, the detection efficiency was determined by Monte Carlo simulation by adding artificial microlensing events (Sumi et al. 2003). They simulated 20 million artificial microlensing events and evaluated the efficiency as a function of Einstein crossing time. The detection efficiency in each field, $\langle \epsilon \rangle$, is given by,

$$\langle \epsilon \rangle = \frac{\sum_i \left[\Gamma(t_{E,i}) \epsilon(t_{E,i}) \right]}{\sum_i \Gamma(t_{E,i})} , \qquad (6.1)$$

where $t_{E,i}$ is the Einstein radius crossing time for the ith event and $\Gamma(t_E)$ is the rate of events of duration t_E for each sub-field. However, in the MOA-II data, detected events are not presented in all sub-fields. Therefore, a Gaussian weighted average of the observed detection efficiency within 1 degree is used to calculate the efficiency. For the all-star sample, the Gaussian weight uses $\sigma = 0.4°$ and $\sigma = 1.0 - 2.5°$ for the RCG sample. Sumi et al. (2013) calculate the microlensing event rate, Γ, from,

$$\Gamma = \frac{1}{N_{\text{source}} T_0} \sum_i \frac{1}{\epsilon(t_{E,i})} , \qquad (6.2)$$

where N_{source} is the total number of source stars and T_0 is the survey duration.

The optical depth, which is the instantaneous number of microlensing events with the impact parameter $u_0 \leq 1$, is calculated using the following equation,

$$\tau = \frac{\pi}{2 N_{\text{source}} T_0} \sum_i \frac{t_{E,i}}{\epsilon(t_{E,i})} . \qquad (6.3)$$

To avoid an optical depth bias towards large t_E events, a maximum event duration $t_{E,\max} = 200$ days is set. The average optical depth in all fields from the MOA-II survey is $1.87^{+0.15}_{-0.13} \times 10^{-6}$ and $1.58^{+0.27}_{-0.23} \times 10^{-6}$ for the all-source sample and RCG sample, respectively. The maps of MOA-II microlensing properties are shown in Fig. 6.2.

The results from the central region ($|l| < 5°$) were binned with $\Delta b = 0.5°$ width and an exponential fit were performed by Sumi et al. (2013). The optical depth as a function of Galactic latitude can be written as

$$\tau_{\text{All-source}} = (2.35 \pm 0.18) \, e^{(0.51 \pm 0.07)(3 - |b|)} \times 10^{-6} ,$$

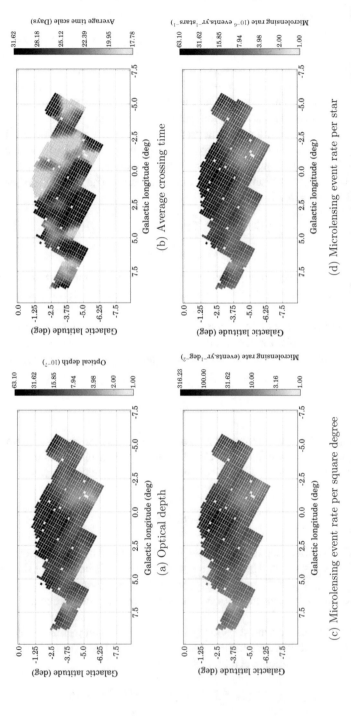

Fig. 6.2 The optical depth (**a**), average time scale (**b**), microlensing event rate per square degree (**c**) and microlensing event rate per star (**d**) from the MOA-II survey. The maps are smoothed by the same kernel function as Sumi et al. (2013)

$$\tau_{RCG} = (1.64 \pm 0.27)\, e^{(0.47\pm0.17)(3-|b|)} \times 10^{-6} \, . \tag{6.4}$$

For the event rate per area, the exponential model is

$$\Gamma_{deg^2, All-source} = (52.03 \pm 2.38)\, e^{(0.50\pm0.04)(3-|b|)} \text{Events.yr}^{-1}\text{deg}^{-2} \, ,$$

$$\Gamma_{deg^2, RCG} = (3.91 \pm 0.48)\, e^{(0.75\pm0.10)(3-|b|)} \text{Events.yr}^{-1}\text{deg}^{-2} \, , \tag{6.5}$$

and for the event rate per star is

$$\Gamma_{star, All-source} = (23.92 \pm 1.13)\, e^{(0.60\pm0.05)(3-|b|)} \times 10^{-6}\text{Events.yr}^{-1}\text{star}^{-1} \, ,$$

$$\Gamma_{star, RCG} = (21.86 \pm 2.64)\, e^{(0.65\pm0.11)(3-|b|)} \times 10^{-6}\text{Events.yr}^{-1}\text{star}^{-1} \, . \tag{6.6}$$

Comparing these results with the theoretical modeling of Kerins et al. (2009), which used the Besançon Galactic model to predict microlensing properties, the observed optical depth contours of the all-sources sample seem to match the model. However, different event selection criteria are used. In this chapter, a later version of Besançon model (Robin et al. 2014) is used to compute MOA-II microlensing property maps, in order to compare observational data with our microlensing model from the latest Besançon Galactic model.

6.2 Simulating the Galactic Bulge of the MOA-II Field

6.2.1 Simulating the MOA-II Fields

We simulate the MOA-II survey data taken from the 2006 and 2007 observing seasons (Sumi et al. 2011, 2013) with the Besançon Galactic model. In order to obtain enough samples in each magnitude range, we produce lens/source star catalogues spanning four H-band magnitude ranges. H-band selection ensures that we adequately sample all relevant stellar types, though we stress that our calculations are performed using the corresponding R and I-band magnitudes of the sources since these are the relevant filters for MOA-II. Our ranges correspond to: $-10 \leq H < 15$, $15 \leq H < 19$, $19 \leq H < 23$ and $H > 23$. The latter ranges are dominated by stars that are too faint to act as sources but do act as lenses. The solid angle in each catalogue, Ω_{sim}, is chosen to contain $\sim$6,000 stars in each range towards Baade's Window ($l = 1°$, $b = -4°$) (Table 6.1). The first catalogue ($-10 \leq H < 15$) has a solid angle of 0.026 deg^2, corresponding to the size of the MOA-II sub-fields. The simulation catalogues stars out to a distance of 15 kpc and has the same overall areal coverage as the MOA-II survey. Our final results are appropriately inverse weighted with Ω_{sim} in order to recover the relevant microlensing observables.

Table 6.1 Solid angles used for the simulated catalogues. Ω_{sim} is used to compute the spatial maps, whilst Ω_{sample} is used to compute the global time scale distribution (Awiphan et al. 2016b)

Magnitude range	Ω_{sim} (deg^2)	Ω_{sample} (deg^2)
$-10 \leq H < 15$	2.6×10^{-2}	4.5×10^{-3}
$15 \leq H < 19$	8.4×10^{-4}	1.4×10^{-4}
$19 \leq H < 23$	9.6×10^{-5}	1.6×10^{-5}
$H > 23$	1.2×10^{-4}	2.0×10^{-5}

For each line of sight, the microlensing optical depth, average time scale and event rate toward the Galactic bulge are calculated using all combinations of source and lens pairs from the four catalogues. We compute microlensing quantities obtained from all resolved sources above a specific magnitude threshold and also from all difference imaging analysis (DIA) sources which have a magnified peak above the same threshold (Alard 2000; Wozniak 2000; Bramich 2008). Therefore, the baseline magnitude of the DIA sources can be fainter than the limit. For unresolved sources, the instantaneous fraction of events with impact parameter u small enough to be detectable scales as u^2, though over time the rate of detectable events scales as u. Therefore, we weight the optical depth by $\min(1, u^2)$ and the rate-weighted average duration by $\min(1, u)$, respectively. The impact parameter moments of Eq. 6.13 are rate-weighted as explained in Sect. 6.3.1 in order to reflect the fact that observables are necessarily obtained from rate-biased samples.

The finite source effect is also taken into account in our calculations even though it contributes a small fraction of the sample. The events that involve a source star with angular radius larger than the angular Einstein radius are not used to calculate the microlensing parameters. However, they are accounted for in the source number normalisation. In practice, this modification alters our results only at the 1% level (see Eq. 6.12).

For sources that have a magnitude fainter than the magnitude limit, the impact parameter is weighted by $\mu D_l R_E$ because of the different frequency of occurrence of microlensing events with different relative proper motion (μ), lens distance (D_l) and Einstein radius (R_E). The weighted average of parameter, $x \leq 1$, can be written as,

$$\langle x \rangle_w = \begin{cases} 1, & M_s < M_{lim} , \\ \dfrac{\sum_{l=1}^{N_l(D_s > D_l)} x \mu D_l R_E}{\sum_{s=1}^{N_s} \sum_{l=1}^{N_l(D_s > D_l)} \mu D_l R_E}, & \text{otherwise} . \end{cases} \tag{6.7}$$

The estimated errors of each quantity are calculated by separating the quantities into two groups and computing their standard deviation. These estimated errors show the variability of quantities from the Besançon Galactic model.

One important issue that we do not explicitly address is source blending. In principle this can be examined within the context of a population synthesis model through construction of artificial images. However, this is beyond the scope of the present work. Instead, we choose to model only the two idealized cases as described in Sect. 5.3 (resolved and DIA sources). If the model is a good representation of the reality, these cases should provide reasonable upper and lower limits on the potential number of events within specific MOA-II sources sub-samples. Our DIA estimate should always provide a firm upper limit to the observed microlensing rate per star. On the other hand, our resolved source calculations should provide a firm lower limit to the number of events per star. MOA-defined sub-samples such as clump giants should yield a rate somewhere intermediate to these regimes as the RCG sources are resolved but known to be confined to the bulge, whereas in our simulation typically more than 15% of our sources are closer to the observer. However, their contribution to the overall optical depth and rate is typically much smaller due to the long disk lens–disk source crossing time. We note that, in order for the model to accurately compute the microlensing rate per unit sky area it would also be necessary to accurately mimic the source colour-magnitude cuts of the survey.

6.2.2   Adding Low-Mass Star, Brown Dwarf and Free Floating Planet Populations

The time scale distribution of the MOA-II observational data, excluding the gb21-R-8-53601 event, which is located outside the Besançon extinction map, and the Besançon simulated data are shown in Fig. 6.3. The histogram of the Besançon data is generated from the sample catalogues using the same criteria as Sect. 6.2.1 but with smaller solid angles, Ω_{sample}, which contain $\sim$1,000 stars at Baade's Window in each catalogue (Table 6.1). From the histogram, the mean crossing time of the Besançon resolved source (25.5 days) and DIA source (26.3 days) samples are larger than the MOA-II mean time scale for all sources (24.0 days) and RGC sources (19.2 days) (Sumi et al. 2013).

In order to investigate the shape of the distributions, the residual event rate distribution ($N'_{Besancon} - N_{MOA}$) is shown in Fig. 6.3, where $N'_{Besancon}$ is Besançon event rate scaled to the number of MOA-II events per year ($\sum N_{MOA}$). The Besançon data shows a deficit of short time scale events ($<$10 days) and an excess of 10–30 day events which may be caused by the lack of low-mass stars and brown dwarfs in the model (Penny et al. 2013). Therefore, we add in low-mass stars and brown dwarf lenses using the same stellar catalogue but replacing the lens mass according to a brown dwarf mass function, as discussed below.

In order to simulate the low mass star population, the same Besançon catalogues from Sect. 6.2.1 are used. In each Galactic component, we add low-mass stars which are missing from the catalogue by extending the normal star mass function slopes, $\alpha \propto \log(dN/dm)/\log m$, to the H-burning limit of $0.079 M_\odot$ which is the minimum mass limit of the Besançon thin disk population (Table 6.2). We assume this limit to be the H-burning limit and use the same limit for the other three populations. We also add in a brown dwarf population with mass function slope, α_{BD}, extending down to 0.001 $M_\odot$ and normalised the number of brown dwarfs to number of stars

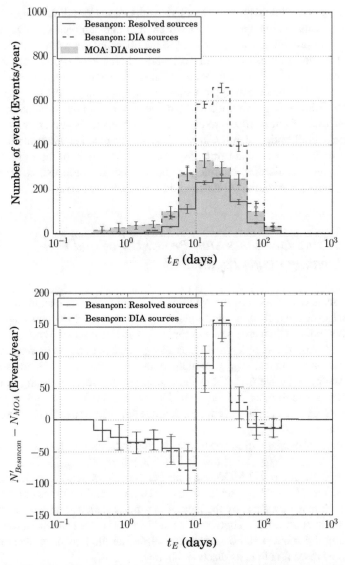

Fig. 6.3 The Einstein radius crossing time distribution of the MOA-II survey and the Besançon data (top) and the residual between the MOA-II survey and the Besançon data, which scaled with the number of MOA-II events, with their MOA-II distribution error (bottom). The blue shaded area (blue thin line) represents the efficiency corrected time scale distribution for the MOA-II DIA sources, excluding event gb21-R-8-53601. The crossing time distribution of the Besançon resolved sources (red thick line) and DIA sources (blue thick dashed line) are also presented. The error bars of the Besançon distributions are shown at 100 times their true size. The red line and blue dashed line represent the residual of the Besançon resolved sources and DIA sources, respectively (Awiphan et al. 2016b)

Table 6.2 The mass function of the simulated low-mass star population (Awiphan et al. 2016b)

Component	Mass range	MF slope
Thick disk	$0.079M_\odot - 0.154M_\odot$	-0.50
Bulge	$0.079M_\odot - 0.150M_\odot$	-1.50
Halo	$0.079M_\odot - 0.085M_\odot$	-0.50

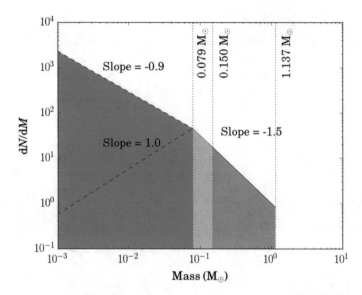

Fig. 6.4 The mass functions of bulge population (Red) and added low-mass stars (Blue) and brown dwarfs (Green)

at the H-burning limit (Fig. 6.4). The additional populations use the same kinetic parameters as the original catalogue and are used for the lens stars only.

In order to find the best value of α_{BD}, the Besançon data from sub-field 7-4 of 20 separate fields (Fields gb1-gb20) normalised by the MOA-II event rate are used to calculate the timescale distributions (Figs. 6.5 and 6.6). In Fig. 6.7, the reduced chi-squares of the predicted versus observed timescale distributions as a function of α_{BD} between -0.9 and 1.0 are shown. Sumi et al. (2011) find that $\alpha_{BD} = -0.49$ for the 2006–2007 MOA-II data. From our simulation, an MF slope of $\alpha_{BD} = -0.4^{+1.9}_{-0.4}$ provides the best reduced chi-square value. This result is consistent with the MOA-II result, but disagrees with the result from some field surveys for young nearby brown dwarfs which suggests a power law MF with slope $\alpha_{bd} > 0.0$ (Kirkpatrick et al. 2012; Jeffries 2012). Therefore, brown dwarfs with mass function slope -0.4 are added to the simulation in order to bring agreement with the observed time scale distribution.

The time scale distribution of the Besançon model with added low-mass stars is shown in Fig. 6.8. The MOA-II survey is analysed using DIA photometry. The number of detected microlensing events per year with efficiency correction from the

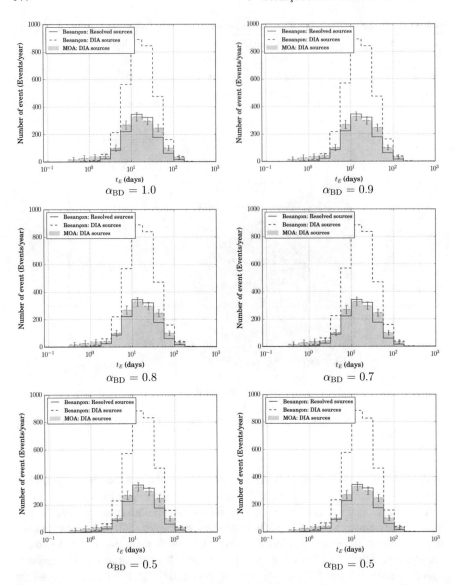

Fig. 6.5 The Einstein radius crossing time distribution of the MOA-II survey and the Besançon data with adding low-mass star and brown dwarfs. The descriptions are the same as in Fig. 6.3

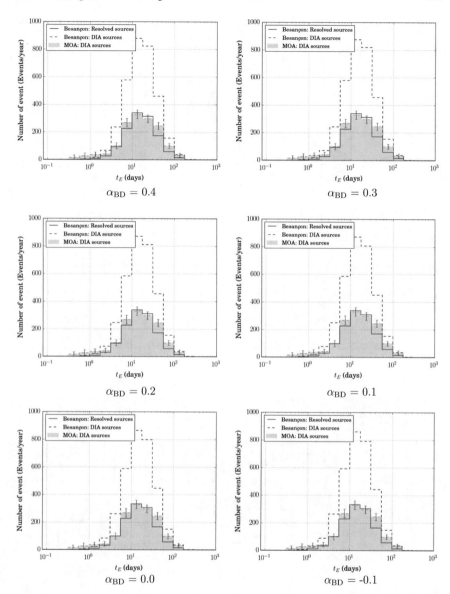

Fig. 6.5 (continued)

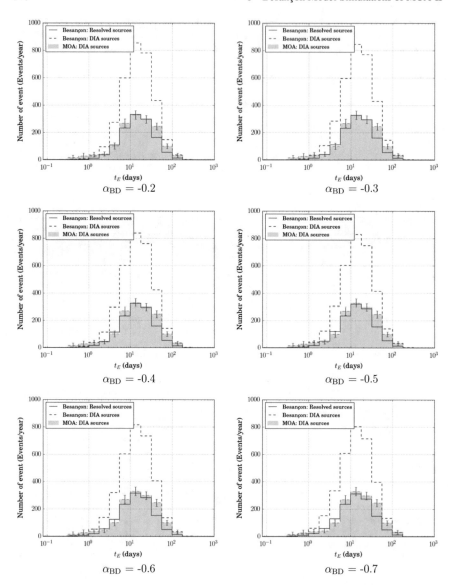

Fig. 6.5 (continued)

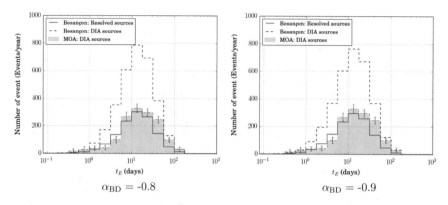

$\alpha_{BD} = -0.8$ $\qquad\qquad\qquad\qquad\qquad\qquad$ $\alpha_{BD} = -0.9$

Fig. 6.5 (continued)

MOA-II survey (N_{MOA}) is between the number of events from the Besançon resolved sources ($0.83\,N_{MOA}$) and DIA sources ($2.17\,N_{MOA}$). In the absence of significant blending effects, we should expect our resolved and DIA predictions to bracket the true result; the fact that it does is rather reassuring. However the effects of blending are complex and a more detailed comparison would require modeling both the source selection criteria and the source blend characteristics of the MOA-II image data. This is beyond the scope of the current work. In the case that all resolved source events are detected, we might be tempted to conclude that 12% of faint stars which can only be detected by the DIA method are observed. However, differences in the assumed filter response can equally be a factor.

The mean crossing times are shorter than the mean crossing time of original Besançon model for both resolved sources and DIA sources, at 20.3 and 20.9 days, respectively. This is close to the MOA-II RCG timescale (19.2 days), but a little lower than their mean timescale for all sources (24.0 days). These mean crossing times also agree with mean crossing times of the OGLE-III survey for resolved sources brighter than $I = 19$ with the relative errors on their crossing time of less than 100%. The average timescales are determined assuming a log-normal model within three regions: positive longitude ($l > 2°$, 22.0 days), central ($-2° < l < 2°$, 20.5 days) and negative longitude ($l < -2°$, 24.2 days) (Fig. 6.8) (Wyrzykowski et al. 2015).

In Fig. 6.7, the residuals of the distribution (model − data) with added low-mass stars show a slight deficit of events with short crossing time between 0.3 and 2 days and very long crossing time between 30 and 200 days. Moreover, the model tends to over-predict the number of events with duration between 2 and 30 days, though there is not a high statistical significance to any of these discrepancies. Overall, our best-fit brown dwarf slope provides a match to the MOA-II timescale distribution with a reduced $\chi^2 \simeq 2.2$ with 19 degrees of freedom (Fig. 6.7). Therefore, the overall mass of the lens population in this simulation is increased about by 10%. The total mass of each population is, $8.0 \times 10^7 M_\odot$ (thin disk), $2.8 \times 10^8 M_\odot$ (thick disk), $4.1 \times 10^8 M_\odot$ (halo) and $5.1 \times 10^8 M_\odot$ (bulge).

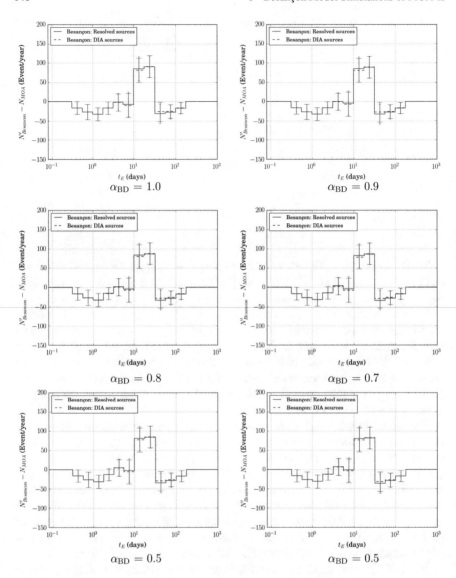

Fig. 6.6 The scaled residual between the MOA-II survey and the Besançon data with added low-mass stars and brown dwarfs with the MOA-II distribution error. The descriptions are the same as in Fig. 6.3

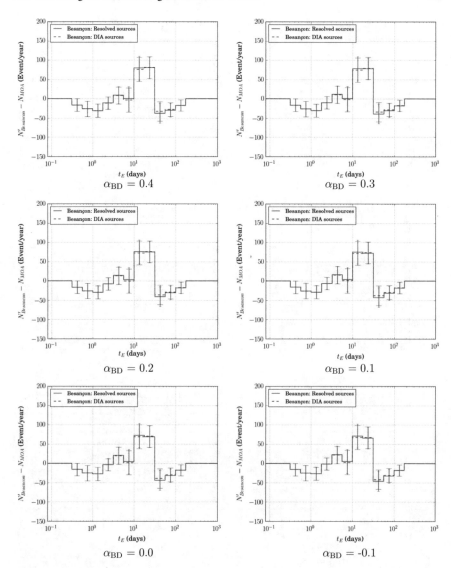

Fig. 6.6 (continued)

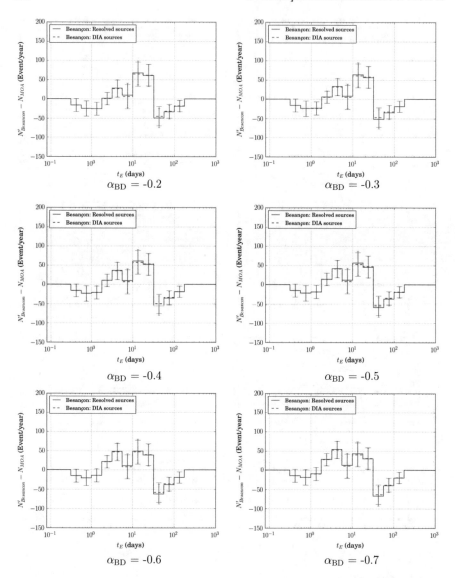

Fig. 6.6 (continued)

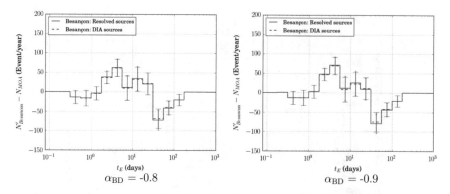

$\alpha_{BD} = -0.8$ $\alpha_{BD} = -0.9$

Fig. 6.6 (continued)

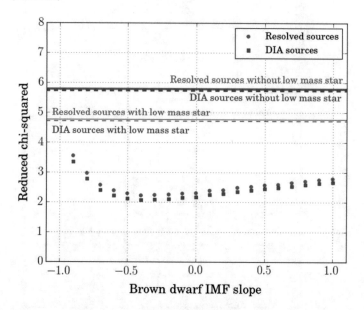

Fig. 6.7 The reduced chi-square values of the model timescale distribution with respect to the MOA-II data for 20 fields is presented as a function of brown dwarf mass function slope. Red circle dots and lines represent the Besançon resolved source data. Blue square dots and dashed lines represent the Besançon DIA source data. Thick lines and thin lines show the original data and the data with added low-mass stars, respectively (Awiphan et al. 2016b)

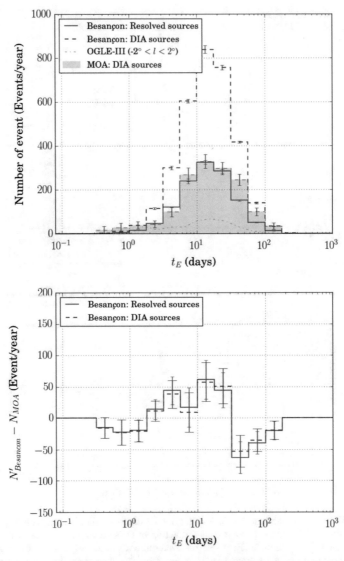

Fig. 6.8 The Einstein radius crossing time distribution of the MOA-II survey, OGLE-III events in $-2° < l < 2°$ fields and the Besançon data with added low-mass stars and brown dwarfs (top). The scaled residual between the MOA-II survey and the Besançon rates (bottom). The descriptions are the same as in Fig. 6.3 (Awiphan et al. 2016b)

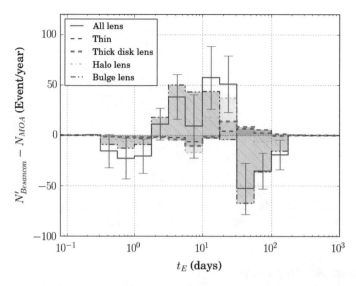

Fig. 6.9 The Einstein radius crossing time residual histogram of the Besançon DIA sources for each Galactic lens component. The residual distribution of all lens components is shown by the red line. The blue thin dashed, green thick dashed, cyan dash-dotted and magenta dashed-dot-dotted lines with shaded areas represent thin disk, thick disk, halo and bulge lenses, respectively (Awiphan et al. 2016b)

To analyse the structure of the residual histogram, we show the contributions of each lens component separately in Fig. 6.9 for DIA sources. The histogram is calculated by assuming that the proportion of each component to the observed rate scales with their proportionate rate within the model. We find that the bulge lens component dominates both the over-predicted and under-predicted regimes, suggesting a mismatch in bulge kinematics, or spatial distribution, as the principal source of the problem.

6.2.3 Timescale Selection

To compare the model optical depth, rate and average duration to the MOA-II survey we must ensure that we match the timescale selection. Accordingly, the maximum Einstein crossing time ($t_{E,\max}$) of this work is set at 200 days to match the MOA-II maximum Einstein crossing time (Sumi et al. 2013). Furthermore, for the minimum Einstein crossing time ($t_{E,\min}$), events with duration below 40 min in fields gb5 and gb9 and 200 min in other fields contribute negligibly (Sumi et al. 2011, 2013). Therefore, the optical depth of all events (τ_{all}) and histograms of Einstein crossing time in each field are used to calculate the optical depth (τ_{select}), average Einstein crossing time ($\langle t_E \rangle_{\text{select}}$) and event rate ($\Gamma_{\text{select}}$) of the events which meet the criteria. Thus

$$\tau_{\text{select}} = \tau_{\text{all}} \frac{\sum_{i=t_{E,\min}}^{t_{E,\max}} t_{E,i}^2 N_i}{\sum_{i=0}^{\infty} t_{E,i}^2 N_i} , \qquad (6.8)$$

$$\langle t_E \rangle_{\text{select}} = \frac{\sum_{i=t_{E,\min}}^{t_{E,\max}} t_{E,i}^2 N_i}{\sum_{i=t_{E,\min}}^{t_{E,\max}} t_{E,i} N_i} , \qquad (6.9)$$

and

$$\Gamma_{\text{select}} = \sum_{i=t_{E,\min}}^{t_{E,\max}} N_i , \qquad (6.10)$$

where $t_{E,i}$ and N_i are the crossing time and the number of microlensing events associated with the *logarithmic* timescale bin i, respectively.

6.3 Microlensing Maps

6.3.1 Optical Depth

Figure 6.10a shows optical depth maps for both resolved and DIA source samples for a survey limit $M_{\lim} = 20$. The maps are computed for the Johnson R and I filter bands, which are comparable to the Cousins R and I bands of the MOA-II survey. The total optical depth of all source and lens pairs is calculated by averaging the optical depth of all sources along the line of sight,

$$\tau = \begin{cases} \dfrac{\sum_{s=1}^{N_s} \sum_{l=1}^{N_l(D_s>D_l, M_s<M_{\lim}, u_p>0)} \dfrac{\pi \theta_E^2}{\Omega_l} \dfrac{\Omega_0}{\Omega_s}}{\sum_{s=1}^{N_s(M_s<M_{\lim})} \dfrac{\Omega_0}{\Omega_s}} & \text{Resolved ,} \\[4ex] \dfrac{\sum_{s=1}^{N_s} \sum_{l=1}^{N_l(D_s>D_l)} u_p^2 \dfrac{\pi \theta_E^2}{\Omega_l} \dfrac{\Omega_0}{\Omega_s}}{\sum_{s=1}^{N_s} \langle u_p^2 \rangle_w \dfrac{\Omega_0}{\Omega_s}} & \text{DIA ,} \end{cases} \qquad (6.11)$$

where M_s is the source magnitude and $M_{\lim}$ is the survey limiting magnitude. Ω_0 is the MOA-II sub-field solid angle. Ω_l and Ω_s are the solid angles over which the lens and source catalogues are simulated, respectively, and N_s and N_l are the number of catalogue sources and lenses, respectively. The impact parameter u_p is given by

$$u_p = \begin{cases} 0, & \theta_E \times \min(1, u_t) \le \theta_* , \\ \min(1, u_t), & \text{otherwise .} \end{cases} \qquad (6.12)$$

Here, θ_* is the source star angular radius and u_t is the largest impact parameter for an event to be detectable above the survey limiting sensitivity. D_s and D_l are the distance to the source and the lens from the observer, respectively. To take account of magnification suppression by finite source size effects, whenever under the point-source regime the source angular size is larger than the largest detectable impact parameter, we assume the event to be undetectable. The n-th moment of u_p, $\langle u_p^n \rangle_w$, is obtained through rate-weighted averaging:

$$\langle u_p^n \rangle_w = \begin{cases} 1, & M_s < M_{\lim} , \\ \dfrac{\sum_{l=1}^{N_l(D_s > D_l)} u_p^n \mu D_l R_E}{\sum_{s=1}^{N_s} \sum_{l=1}^{N_l(D_s > D_l)} \mu D_l R_E}, & \text{otherwise} , \end{cases} \tag{6.13}$$

where μ is the lens–source pair-wise relative proper motion.

We employ the same Gaussian spatial smoothing window function as Sumi et al. (2013), with $\sigma = 0.4°$ within $1°$. In order to compare the data with the MOA-II results, the sub-fields with $l > 9°$ are excluded due to the kernel contribution from sub-fields outside the Besançon extinction map at $l > 10°$.

From the simulation results, the optical depth of DIA sources is larger than the optical depth of resolved sources, as was also found by Kerins et al. (2009) using an earlier version of the Besançon model. However the current model predicts a significantly lower optical depth compared with Kerins et al. (2009) due to the lower mass of the Galactic bulge (which is a factor of two lower than for the earlier model). However, the optical depth values are compatible with the Penny et al. (2013) result which also uses a more recent version of the Besançon model (version 1106). In Figs. 6.11 and 6.12, the DIA sources provide a higher optical depth compared to the resolved sources as expected. The optical depth distribution is dominated by the bulge population which contains about 50–80% of the lens stars. The thin disk, thick disk and stellar halo lenses provide slightly larger optical depth contributions at negative longitude than positive longitude due to the fact that bulge sources, which dominate the statistics, tend to lie at larger distances at negative longitudes.

6.3.2 Maps of Average Event Duration

In order to calculate the average time scale, each Einstein crossing time (t_E) is rate-weighted by a factor $\mu D_l R_E$. Finally, the average time scale, $\langle t_E \rangle$, is obtained as

$$\langle t_E \rangle = \begin{cases} \dfrac{\sum_{s=1}^{N_s} \sum_{l=1}^{N_l(D_s > D_l, M_s < M_{\lim}, u_p > 0)} \theta_E^2 D_l^2 \frac{\Omega_0}{\Omega_l \Omega_s}}{\sum_{s=1}^{N_s} \sum_{l=1}^{N_l(D_s > D_l, M_s < M_{\lim})} \theta_E D_l^2 \mu \frac{\Omega_0}{\Omega_l \Omega_s}} & \\ \hspace{4cm} \text{(Resolved)}, & \\[2em] \dfrac{\sum_{s=1}^{N_s} \sum_{l=1}^{N_l(D_s > D_l)} u_p \theta_E^2 D_l^2 \frac{\Omega_0}{\Omega_l \Omega_s}}{\sum_{s=1}^{N_s} \sum_{l=1}^{N_l(D_s > D_l)} \langle u_p \rangle_w \theta_E D_l^2 \mu \frac{\Omega_0}{\Omega_l \Omega_s}} & \\ \hspace{4cm} \text{(DIA)}, & \end{cases} \tag{6.14}$$

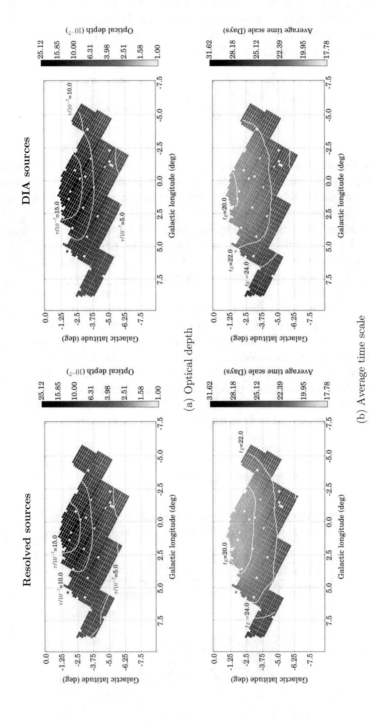

Fig. 6.10 The optical depth (**a**), average time scale (**b**), microlensing event rate per square degree (**c**) and microlensing event rate per star (**d**) for resolved sources (left) and DIA sources (right) from the Besançon Galactic model. The maps are smoothed by the same kernel function as Sumi et al. (2013). The sub-fields with $l > 9°$ are excluded (Awiphan et al. 2016b)

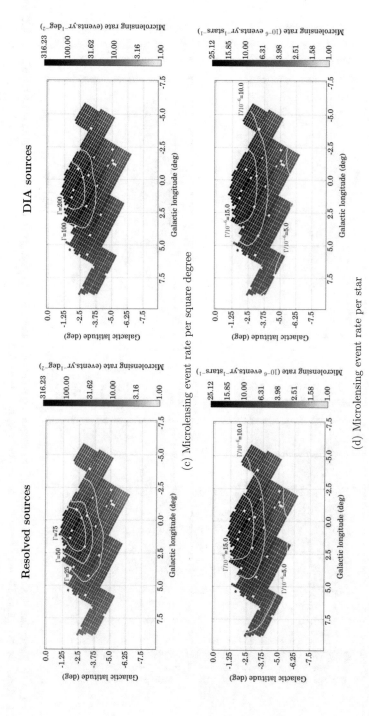

(c) Microlensing event rate per square degree

(d) Microlensing event rate per star

Fig. 6.10 (continued)

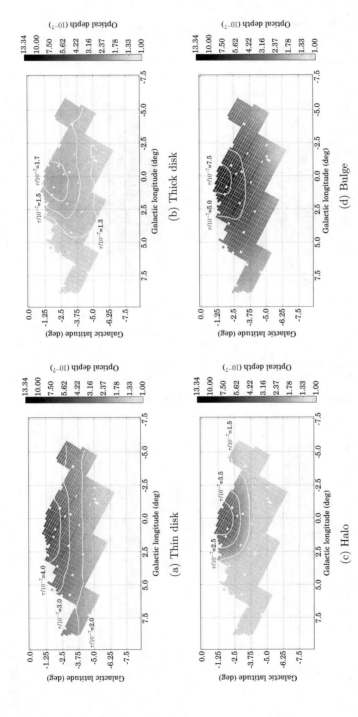

Fig. 6.11 The optical depth maps of each Galactic lens component: **a** thin disk lenses, **b** thick disk lenses, **c** stellar halo lenses and **d** bulge lenses with the resolved sources method. The maps have same description as the map in Fig. 6.10

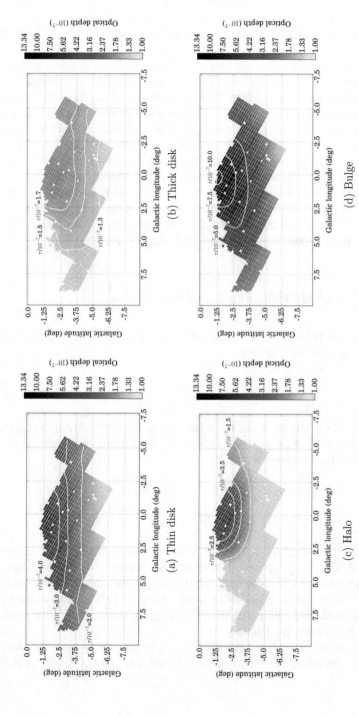

Fig. 6.12 The optical depth maps of each Galactic lens component: **a** thin disk lenses, **b** thick disk lenses, **c** stellar halo lenses and **d** bulge lenses with the DIA method. The maps have same description as the map in Fig. 6.10 (Awiphan et al. 2016b)

Maps of the average event duration are shown in Fig. 6.10b. The maps show shorter timescales compared to Kerins et al. (2009) and Penny et al. (2013), due to the addition of low-mass star and brown dwarf lenses. There is no major difference between the average time scale of resolved sources and DIA sources. The negative longitudes provide slightly longer time scales than positive longitudes due to the bar geometry resulting in typically larger Einstein radii at negative longitudes.

In Figs. 6.13 and 6.14, we show the average timescale maps individually for the thin disk, thick disk, stellar halo lens and bulge lens populations for resolved sources and DIA sources, respectively. The maps show a reasonably symmetric spatial distribution in the average event duration, with bulge lenses exhibiting typically shorter time scales compared to other lens components. Since bulge lenses dominate the event rate in the inner Galaxy (Figs. 6.15 and 6.16) the overall map of event duration shown in Fig. 6.10b closely resembles that of the bulge lens population. We also confirm from Figs. 6.13 and 6.14 that the long duration region at longitude $l > 7.5°$ evident in Fig. 6.10b arises from the disk lens population as the density of bulge lenses become sub-dominant away from the Galactic Centre.

6.3.3   Map of Microlensing Event Rate

The total event rate is obtained by dividing the optical depth maps by their corresponding average time scale maps:

$$\Gamma = \frac{2}{\pi} \frac{\tau}{\langle t_E \rangle} . \tag{6.15}$$

Figure 6.10c, d show maps of microlensing event rate per square degree (Γ_{deg^2}) and event rate per star (Γ_{star}), respectively. Γ_{deg^2} is obtained by integrating the rate over the effective number of sources:

$$N = \begin{cases} \sum_{s=1}^{N_s(M_s > M_{\mathrm{lim}})} \frac{\Omega_0}{\Omega_s} & \text{Resolved} , \\ \sum_{s=1}^{N_s} \langle u_p \rangle_w \frac{\Omega_0}{\Omega_s} & \text{DIA}. \end{cases} \tag{6.16}$$

In Fig. 6.10c we see that Γ_{deg^2} for DIA sources is higher than for resolved sources, as expected. The area integrated microlensing event rate in the simulated maps for resolved sources and DIA sources is 1,250 and 3,250 events per year, respectively. The maps of Γ_{star} in Fig. 6.10d for resolved sources and DIA sources do not show a major difference indicating that, overall, they probe sources and lenses at similar distances with similar kinematics.

In Figs. 6.15 and 6.16, the maps of Γ_{star} are shown separately for each lens population. The strong dominance of bulge lenses over most of the MOA-II region is evident.

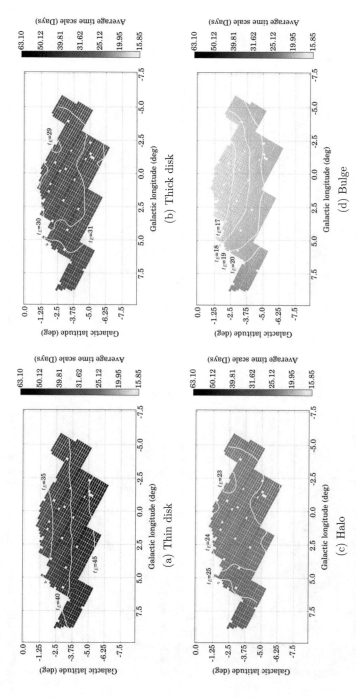

Fig. 6.13 The average time scale maps of each Galactic lens component for resolved sources: **a** thin disk, **b** thick disk, **c** halo and **d** bulge. The maps have same description as the map in Fig. 6.10

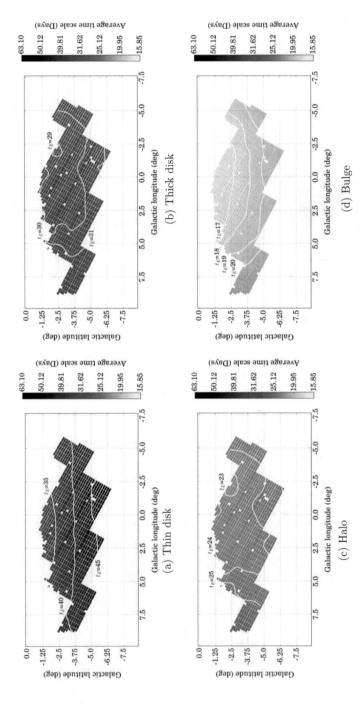

Fig. 6.14 The average time scale maps of each Galactic lens component for DIA sources: **a** thin disk, **b** thick disk, **c** halo and **d** bulge. The maps have same description as the map in Fig. 6.10 (Awiphan et al. 2016b)

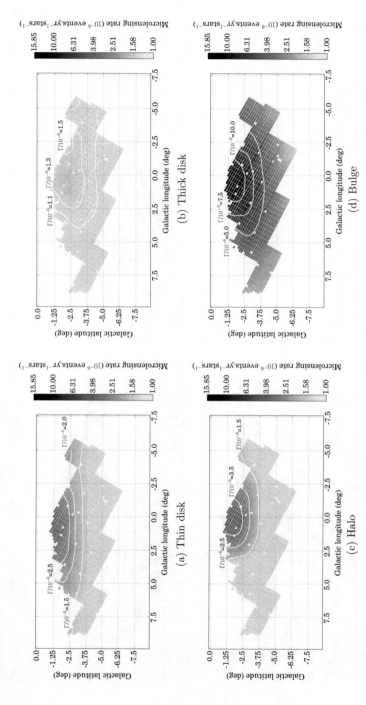

Fig. 6.15 The microlensing event rate per star maps of each Galactic lens component for resolved sources: **a** thin disk, **b** thick disk, **c** halo and **d** bulge. The maps have same description as the map in Fig. 6.10

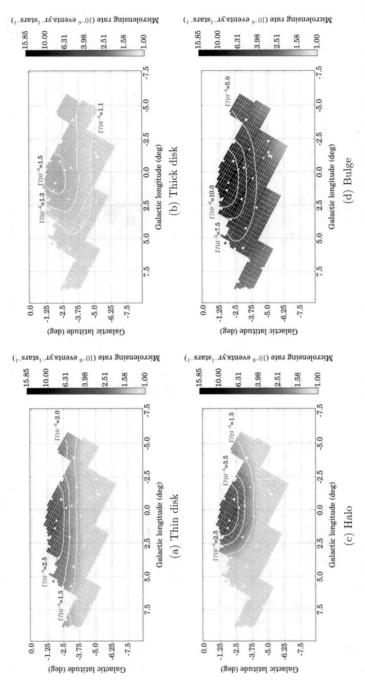

Fig. 6.16 The microlensing event rate per star maps of each Galactic lens component for DIA sources: **a** thin disk, **b** thick disk, **c** halo and **d** bulge. The maps have same description as the map in Fig. 6.10 (Awiphan et al. 2016b)

6.4 Confronting MOA-II Observations

6.4.1 Galactic Latitude Variation

The optical depth, microlensing event rate per square degree and microlensing event rate per star from the Besançon Galactic model and survey observations at different Galactic latitudes are presented in Fig. 6.17. The results are calculated from the optical depth and microlensing event rate between $l = -5°$ and $l = 5°$. The stars in each simulated sub-field are binned to $0.5°$ in Galactic latitude, in similar fashion to the MOA-II survey (Sumi et al. 2013). The results from the previous measurements, as well as the simulated models of this work, are listed in Tables 5.1 and 6.3, respectively. The shaded areas in Fig. 6.17 show the 68% confidence interval of the data. The shaded 68% confidence intervals are obtained by generating random deviate distributions of each exponential fit model assuming that the errors on the best-fit parameters are Gaussian distributed.

From Fig. 6.17a comparing the MOA-II data and the Besançon simulated data, the optical depth at $b > 1.5°$ agrees very well with an exponential fit. For $b < 1.8°$, the optical depths decrease due to the high column density of dust in that area. Over the lower latitude regions ($b < 3°$), the Besançon DIA optical depth is lower than the MOA-II all-source optical depth by a factor of 2, a factor similar to that found by Penny et al. (2013) (Fig. 6.18). The exponential models of the Besançon optical depth are,

$$\tau_{Res} = (1.18 \pm 0.03)\, e^{(0.330 \pm 0.017)(3-|b|)} \times 10^{-6}\,,$$

$$\tau_{DIA} = (1.31 \pm 0.02)\, e^{(0.357 \pm 0.013)(3-|b|)} \times 10^{-6}\,. \qquad (6.17)$$

In Fig. 6.17b, the event rate per square degree for the Besançon resolved sources is compatible with the MOA-II all-source event rate, however for DIA sources it is 3 times higher than the MOA-II result. This might be a consequence of the blending effect as discussed in Sect. 6.2.2. The results from both also show the same turning point at $l = 1.8°$ as the optical depth. The exponential fits for the event rate per square degree give,

$$\Gamma_{deg^2,Res} = (43 \pm 4)\, e^{0.380 \pm 0.510)(3-|b|)} \text{Events.yr}^{-1}\text{deg}^{-2}\,,$$

$$\Gamma_{deg^2,DIA} = (119 \pm 9)\, e^{(0.510 \pm 0.060)(3-|b|)} \text{Events.yr}^{-1}\text{deg}^{-2}\,. \qquad (6.18)$$

For the microlensing event rate per star, Fig. 6.17c, the results are similar to the optical depth result, which is expected given the general agreement of the average event duration. The exponential models of the simulated event rate per star can be written as,

$$\Gamma_{star,Res} = (13.5 \pm 0.4)\, e^{(0.362 \pm 0.023)(3-|b|)} \times 10^{-6} \text{Events.yr}^{-1}\text{star}^{-1}\,,$$

Fig. 6.17 The optical depth (**a**), microlensing event rate per square degree (**b**) and microlensing event rate per star (**c**) as a function of Galactic latitude. The measurements are averaged over Galactic longitudes $-5° < l < 5°$. Different markers represent different survey measurements (See Table 5.1): OGLE (pentagon), MACHO (circle), MOA (triangle), EROS (square) and simulated data from the Besançon Galactic model (star) (See Table 6.3). Results of resolved sources, DIA sources and RCG source are presented with unfilled, filled and half-filled makers. The error bars of the Besançon simulation results are shown at 100 times their true size. The thin dashed, dash-dotted and dotted lines represent fits to the MOA-II all-source sample, EROS RCG sample and MOA-II RCG sample, respectively (Hamadache et al. 2006; Sumi et al. 2013). The thick solid and dashed lines are fits to the resolved source and DIA source simulations of this work. The shaded areas represent the 68% confidence interval of EROS, MOA-II and Besançon fits, respectively (Awiphan et al. 2016b)

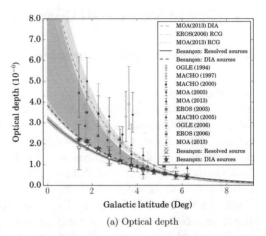

(a) Optical depth

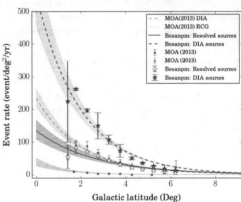

(b) Microlensing event rate per square degree

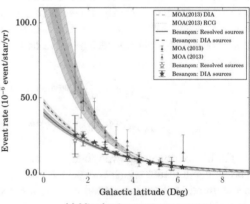

(c) Microlensing event rate per star

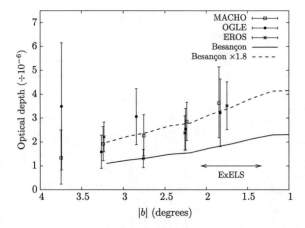

Fig. 6.18 A comparison of measured optical depths to red clump giants from the Besançon model from Penny et al. (2013). The open square, filled circle and asterisk data points show the measurement from the MACHO (Popowski et al. 2005), OGLE Sumi et al. (2006) and EROS Hamadache et al. (2006) surveys, respectively. The solid line shows the optical depth from the Besançon model. The dashed line shows 1.8 times of the solid curve (Penny et al. 2013)

$$\Gamma_{star,DIA} = (14.6 \pm 0.3)\, e^{(0.391 \pm 0.016)(3 - |b|)} \times 10^{-6} \text{Events.yr}^{-1} \text{star}^{-1} \, . \qquad (6.19)$$

Whilst the model average event duration is in reasonable agreement with the MOA-II observations, the factor of 2 discrepancy with both the optical depth and rate suggests that the model bulge mass is too low to accommodate the microlensing results. The bulge mass would need to be increased by a factor of 2.6 in order to match the overall optical depth distribution. However, we note that such a change could not be made in a self-consistent manner without also altering the lens and source kinematics and the event timescale distribution. However, in Sect. 6.6, we discuss an updated MOA-II result.

6.4.2 O-C Residual Maps

In Sect. 6.3, the model microlensing maps of the MOA-II field, filter and time scale cut are shown. In order to compare the result with the MOA-II observational data, residual maps are produced ($x_{Besancon} - x_{MOA}$). In Fig. 6.19a, the Besançon model underestimates the optical depth compared with the MOA-II data closer to the Galactic Centre ($b < 3°$), predicting only around 50% of the MOA-II measured optical depth. However, moving away from the Galactic Centre, the Besançon optical depth is generally in good agreement with the MOA-II measurement, suggesting that the Besançon disk model provides a good description of the microlensing data.

Table 6.3 The Besançon model microlensing optical depth and event rate within $|l| < 5°$ using 0.5° bins in b (Awiphan et al. 2016b)

$\langle b \rangle$	N_{event} (events)	$\tau \times 10^7$	Γ_{deg^2} $(yr^{-1}deg^{-2})$	Γstar $(\times 10^{-6}yr^{-1}star^{-1})$
Resolved source				
$-1.40°$	18.414	18.414 ± 0.109	54.828 ± 0.347	21.774 ± 0.111
$-1.77°$	18.759	18.759 ± 0.017	83.258 ± 0.183	22.671 ± 0.048
$-2.26°$	16.123	16.123 ± 0.006	70.568 ± 0.009	19.209 ± 0.009
$-2.76°$	13.311	13.311 ± 0.035	58.620 ± 0.092	15.551 ± 0.034
$-3.25°$	11.038	11.038 ± 0.007	43.998 ± 0.012	12.670 ± 0.003
$-3.75°$	9.055	9.055 ± 0.003	30.954 ± 0.043	10.162 ± 0.014
$-4.25°$	7.494	7.494 ± 0.023	22.192 ± 0.034	8.146 ± 0.015
$-4.74°$	6.295	6.295 ± 0.010	16.486 ± 0.011	6.646 ± 0.003
$-5.23°$	5.488	5.488 ± 0.002	12.685 ± 0.036	5.632 ± 0.017
$-5.72°$	4.589	4.589 ± 0.003	8.919 ± 0.058	4.601 ± 0.030
$-6.23°$	3.934	3.934 ± 0.009	6.328 ± 0.007	3.809 ± 0.004
DIA source				
$-1.40°$	21.931	21.931 ± 0.006	224.902 ± 1.242	25.553 ± 0.129
$-1.77°$	21.301	21.301 ± 0.009	261.791 ± 0.031	24.804 ± 0.014
$-2.26°$	17.846	17.846 ± 0.001	197.174 ± 0.049	20.509 ± 0.003
$-2.76°$	14.597	14.597 ± 0.040	150.158 ± 0.396	16.489 ± 0.040
$-3.25°$	12.024	12.024 ± 0.003	107.772 ± 0.135	13.364 ± 0.012
$-3.75°$	9.816	9.816 ± 0.030	73.464 ± 0.198	10.689 ± 0.028
$-4.25°$	8.054	8.054 ± 0.004	50.647 ± 0.039	8.542 ± 0.006
$-4.74°$	6.730	6.730 ± 0.001	36.212 ± 0.030	6.928 ± 0.006
$-5.23°$	5.786	5.786 ± 0.005	27.041 ± 0.007	5.819 ± 0.003
$-5.72°$	4.848	4.848 ± 0.019	18.826 ± 0.100	4.771 ± 0.027
$-6.23°$	4.153	4.153 ± 0.014	13.182 ± 0.010	3.958 ± 0.003

The residual average time scale maps in Fig. 6.19b, show that the Besançon model provides a reasonably good representation of the MOA-II timescales across the whole map. Most of the structures in the map are produced by individual very long time scale MOA-II events or many short time scale events in observed specific sub-fields. We therefore conclude that the *average* microlensing kinematics within both the disk and bulge are consistent with microlensing data.

Figure 6.19c, d, which show the residual event rate per unit area and per source, respectively, indicate a similar deficit of the model with respect to the data within the inner bulge region.

These maps confirm the view that, whilst the model bulge kinematics provide a good average description of the event timescale, the mass is insufficient by a factor 2 to explain the observed number of events. However, recently the MOA-II measurement has been revised (Sumi and Penny 2016) and we discuss the consequences of the is Sect. 6.6.

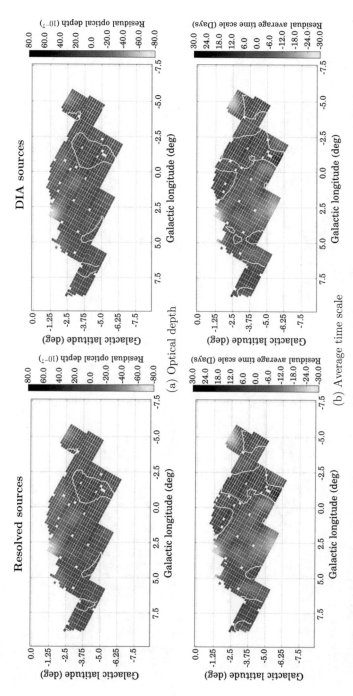

Fig. 6.19 The optical depth (**a**), average time scale (**b**), microlensing event rate per square degree (**c**) and microlensing event rate per star (**d**) residual maps of resolved sources (left) and DIA sources (right) from the Besançon Galactic model and the MOA-II survey data. Contour level shows zero residual value (Awiphan et al. 2016b)

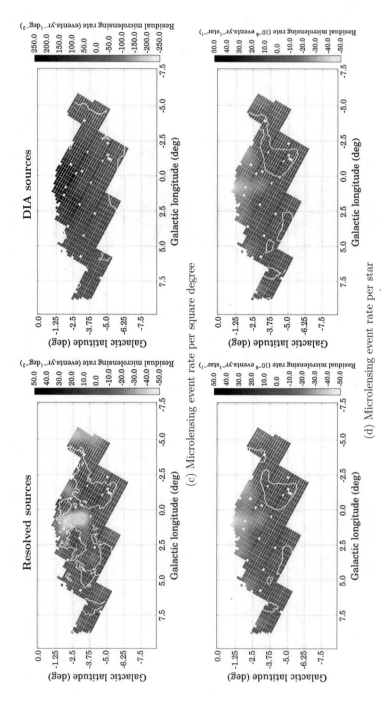

(c) Microlensing event rate per square degree

(d) Microlensing event rate per star

Fig. 6.19 (continued)

6.4.3 Reduced χ^2 of τ and Γ_{star} Maps

From Sect. 6.4.2, the simulated results under-predict the optical depth and microlensing event rate per star compared with the MOA-II observational data and show the structure at low Galactic latitude. The significance of this result can be assessed by a straightforward reduced χ^2 statistic:

$$\chi_r^2 = \frac{1}{N_{fld}} \sum_{i=1}^{N_{fld}} \left(\frac{x_{Bes,i} - x_{MOA,i}}{\sigma_{MOA,i}} \right)^2, \qquad (6.20)$$

where (x_{Bes}, x_{MOA}) refers to the (model, observed) microlensing quantity. σ is the observational uncertainty within the field, and N_{fld} is number of fields. The observational uncertainty within the field is calculated using the formula from Han and Gould (1995). The reduced χ^2 contribution of each MOA-II field are shown in Table 6.4 and Fig. 6.20. The gb21 field is excluded due to the limit of the Besançon extinction maps. The model optical depth is in agreement with MOA-II data within $3\sigma_{MOA}$ for most fields. The reduced χ^2 of resolved source and DIA source optical depths are 2.4 and 2.0, respectively.

The event rates show higher reduced chi-squared contribution than the optical depths. The Besançon resolved source and DIA source results have χ_r^2 values of 2.6 and 2.2, respectively. The low Galactic latitude area ($b < 3°$) of both parameters provide the bulk of the disagreement (See Sect. 6.4.1).

In field gb1, there is a long time scale event (gb1-3-1, $t_E = 157.6$ days) which contributes more than half of optical depth in that field. This event provides a hot spot in the MOA-II optical depth and average time scale maps (See Sect. 6.4.2). In order to check the reliability of the reduced χ^2 test, the reduced χ^2 of field gb1 without gb1-3-1, gb1$_{Cut}$, is calculated. The result in Table 6.4 shows that the gb1$_{Cut}$ field provides a better reduced χ_r^2 than original gb1 field.

Finally, we cut the events which have crossing time longer than 100 days which are located in 5 fields; gb1, gb9, gb10, gb13 and gb 14. The new reduced chi-squared, $\chi_{r,Cut}^2$, of optical depth (2.3 for resolved sources and 1.8 for DIA sources) and event rate per star (2.2 for resolved sources and 2.0 for DIA sources) is improved. However, fields located in the inner bulge, except field gb6, still show high reduced chi-squared compared to high Galactic latitude field. Therefore, the high reduced chi-squared region around the inner bulge is not affected by long time scale events, but shows the mismatch of optical depth and event rate per star between the Besançon model and the MOA-II data in low Galactic latitude region.

6.5 Microlensing Model Parametrisation

The MOA-II team parameterises the observed spatial microlensing distribution using a polynomial function. We can do likewise for our simulated maps. We model the

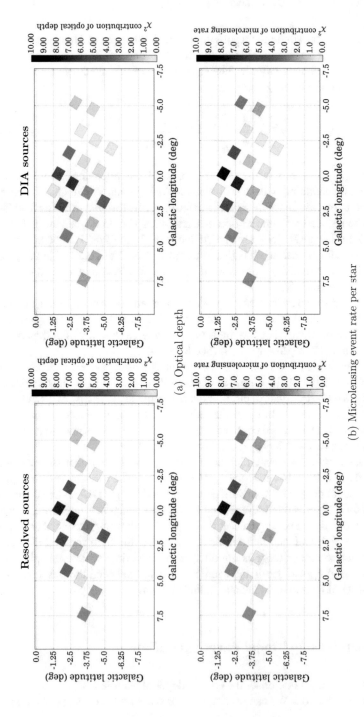

Fig. 6.20 The reduced χ^2 contribution of optical depth (**a**) and microlensing event rate per star (**b**) of resolved sources (left) and DIA sources (right) (Awiphan et al. 2016b)

Table 6.4 Field-by-field contributions to the reduced χ^2 (χ^2_r) between the Besançon model and the MOA-II data for the optical depth and microlensing event rate per star (Awiphan et al. 2016b)

Field	$\langle l \rangle$ (deg)	$\langle b \rangle$ (deg)	τ_{Res}	τ_{DIA}	$\Gamma_{star,Res}$	$\Gamma_{star,DIA}$
gb1	$-4.3306°$	$-3.1119°$	1.40	1.22	3.67	3.00
gb2	$-3.8624°$	$-4.3936°$	1.34	0.91	2.47	2.09
gb3	$-2.3463°$	$-3.5133°$	1.01	0.45	0.02	0.06
gb4	$-0.8210°$	$-2.6317°$	4.31	3.24	5.91	4.78
gb5	$0.6544°$	$-1.8595°$	7.31	5.47	9.41	7.94
gb6	$1.8405°$	$-1.4890°$	0.04	0.01	0.15	0.02
gb7	$-1.7147°$	$-4.5992°$	0.28	0.08	0.67	0.44
gb8	$-0.1937°$	$-3.7495°$	0.26	0.77	0.84	1.44
gb9	$1.3329°$	$-2.8786°$	7.34	6.35	7.99	7.08
gb10	$2.8448°$	$-2.0903°$	5.14	4.56	5.76	4.91
gb11	$-1.1093°$	$-5.7257°$	10^{-3}	0.02	0.38	0.29
gb12	$0.4391°$	$-4.8658°$	0.97	0.75	0.03	10^{-3}
gb13	$1.9751°$	$-4.0190°$	2.77	2.50	2.03	1.76
gb14	$3.5083°$	$-3.1698°$	1.72	1.39	1.74	1.34
gb15	$4.9940°$	$-2.4496°$	3.37	2.75	3.12	2.60
gb16	$2.6048°$	$-5.1681°$	3.98	3.73	2.36	2.13
gb17	$4.1498°$	$-4.3365°$	1.80	1.53	0.45	0.32
gb18	$5.6867°$	$-3.5055°$	0.64	0.32	0.27	0.12
gb19	$6.5534°$	$-4.5749°$	2.12	1.77	1.28	1.02
gb20	$9.6172°$	$-2.9318°$	2.56	2.02	2.91	2.49
χ^2_r			2.4	2.0	2.6	2.2
gb1$_{Cut}$	$-4.3306°$	$-3.1119°$	0.87	0.33	2.72	2.10
gb9$_{Cut}$	$1.3329°$	$-2.8786°$	7.34	6.34	7.72	6.83
gb10$_{Cut}$	$2.8448°$	$-2.0903°$	4.20	3.34	4.35	3.57
gb13$_{Cut}$	$1.9751°$	$-4.0190°$	1.89	1.57	1.50	1.26
gb14$_{Cut}$	$3.5083°$	$-3.1698°$	0.77	0.40	1.05	0.72
$\chi^2_{r,Cut}$			2.3	1.8	2.4	2.0

structure of the optical depth, average time scale and event rate maps shown in Fig. 6.10 using a 10-parameter cubic polynomial fit in l and b. The model function can be written as,

$$x = a_0 + a_1 l + a_2 b + a_3 l^2 + a_4 lb + a_5 b^2 + a_6 l^3 + a_7 l^2 b + a_8 lb^2 + a_9 b^3 ,$$
(6.21)

where x is the microlensing observable (rate, time-scale or optical depth). The best-fit models are shown in Fig. 6.21 and the model parameters are provided in Table 6.5. The best fit models agree to within 20% of the exact model value for $|b| < 5°$.

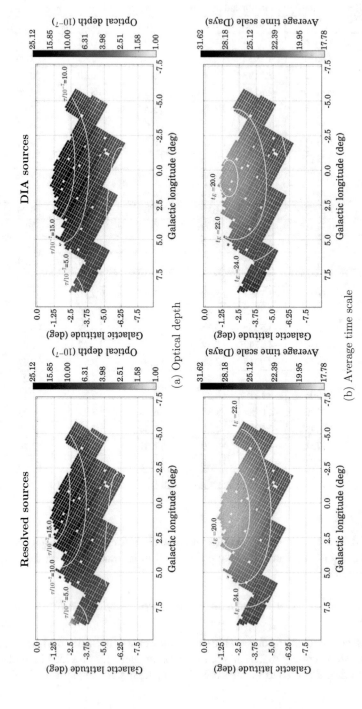

Fig. 6.21 Best-fit parametrised representations of the Besançon model maps shown in Fig. 6.10. Best-fit parametrisations are shown for the optical depth (**a**), average time scale (**b**), microlensing event rate per square degree (**c**) and microlensing event rate per star (**d**) for resolved sources (left) and DIA sources (right). The parameters of the fits are given in Table 6.5 (Awiphan et al. 2016b)

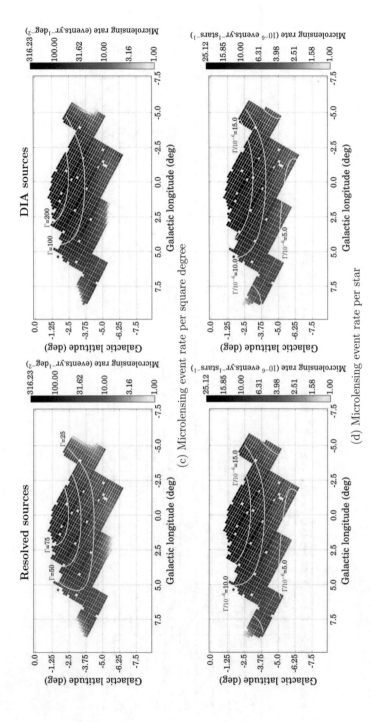

(c) Microlensing event rate per square degree

(d) Microlensing event rate per star

Fig. 6.21 (continued)

Table 6.5 The best fit model parameters of the Besançon microlensing model (Awiphan et al. 2016b)

	τ_{Res} $\times 10^{-8}$	τ_{DIA} $\times 10^{-8}$	$\langle t \rangle_{Res}$ days	$\langle t \rangle_{DIA}$ days	$\Gamma_{deg^2,Res}$ $yr^{-1}deg^{-2}$	$\Gamma_{deg^2,DIA}$ $yr^{-1}deg^{-2}$	$\Gamma_{star,Res}$ $\times 10^{-7}yr^{-1}star^{-1}$	$\Gamma_{star,DIA}$ $\times 10^{-7}yr^{-1}star^{-1}$
a_0	256	321	21.7	21.5	101	477	305	378
a_1	−14.8	−16.9	−0.151	−0.260	−11.0	−38.9	−17.2	−19.0
a_2	35.8	63.5	2.63	1.91	−14.5	111	38.4	75.1
a_3	−4.21	−4.52	0.121	0.126	−3.99	−10.3	−5.63	−5.89
a_4	−7.85	−8.70	0.012	−0.043	−7.49	−22.9	−9.42	−10.1
a_5	−3.84	0.464	0.817	0.631	−12.9	−1.02	−6.93	−0.191
a_6	0.087	0.104	0.002	0.003	0.053	0.191	0.112	0.127
a_7	−0.720	−0.771	0.016	0.016	−0.735	−1.88	−0.965	−1.00
a_8	−1.02	−1.12	0.004	−0.004	−1.08	−3.13	−1.25	−1.32
a_9	−0.647	−0.408	0.056	0.042	−1.31	−1.10	−1.00	−0.570

6.6 Comparing with MOA-II Corrected Stellar Number Count Data

Sumi and Penny (2016) pointed out that the number of sources in Sumi et al. (2013) used the luminosity at Baade's window for all fields, which may cause incompleteness in star counts. This incompleteness affects the measured optical depth and event rate for MOA-II data. Therefore, Sumi and Penny (2016) revised the measurement of the MOA-II optical depth and event rate with completed star counts of sources using the same data set as Sumi et al. (2013).

Sumi et al. (2013) used the star catalogues at Baade's window from the MOA-II reference image for bright stars and *Hubble* Space Telescope images for faint stars (Holtzman et al. 1998) to derive the luminosity function using the method of Nataf et al. (2013). They normalized the luminosity function for each sub-field. Therefore, the shape of the luminosity function in all fields is the same as the function at Baade's window. However, after Sumi and Penny (2016) revised the data using the new luminosity function, they found that the effects are negligible.

The event selection criteria of Sumi et al. (2013) selected the event regardless of whether the sources are resolved stars. Their star counts were based on stellar catalogue in the reference images, in which red clump giants were assumed to be bright enough for complete star counts. However, the MOA-II data were obtained under poor seeing ($\sim$1.8 arcsec), including the reference image. Therefore, the red clump giant number counts are incomplete.

The incompleteness can be proven by comparing the number of red clump giants per sub-field in MOA-II and OGLE which has better seeing and longer exposure than MOA-II. The OGLE data used a different method for event selection, which selected only the events at the position of the resolved stars in the reference images. For the low number of red clump giants ($N_{RC} < 1,000$), the number of red clump giants in MOA-II are consistent with OGLE. But, the number of RCG stars in MOA-II is smaller than OGLE by up to 30% in higher density region. The number of missing stars in the MOA-II data is expected to be a few percent. Sumi and Penny (2016) also found that the most reliable data are located at high Galactic latitude ($b \sim -6$) and are faint sources. These properties are related to the Nataf et al. (2013) method. Therefore, Sumi and Penny (2016) fit relations between the number of red clump giants in MOA-II, $N_{RC,MOA}$, and the Nataf et al. (2013) predicted number of red clump giants, $N_{RC,Nataf}$, as a function of Galactic latitude.

$$\frac{N_{RC,MOA}}{N_{RC,Nataf}} = (0.63 \pm 0.01) - (0.052 \pm 0.003) \times b . \tag{6.22}$$

The mismatch of star counts affected the optical depth and event rate measurement of Sumi et al. (2013). Sumi and Penny (2016) revised the optical depth and event rate measurements using the relation in Eq. 6.22, however the event rate per square degree does not change from previous measurements due to it being independent of the number of stars. Therefore, the optical depth and microlensing event rate per star

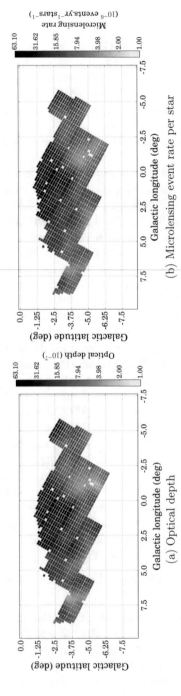

Fig. 6.22 The optical depth (**a**) and microlensing event rate per star (**b**) from the revised MOA-II data (Sumi and Penny 2016)

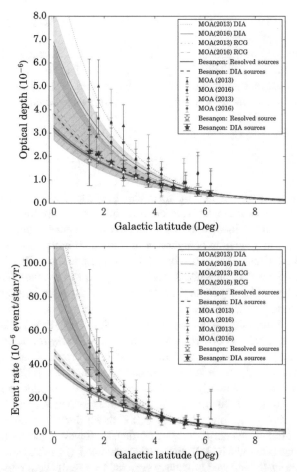

Fig. 6.23 The optical depth (top) and microlensing event rate per star (bottom) as a function of Galactic latitude of the revised MOA-II data. The measurements are averaged over Galactic longitudes $-5° < l < 5°$. Different markers represent different survey measurements: MOA Sumi et al. (2013) (triangle), MOA Sumi and Penny (2016) (circle) and simulated data from the Besançon Galactic model (star) (See Table 6.3). Results of resolved sources, DIA sources and RCG source are presented with unfilled, filled and half-filled makers. The error bars of the Besançon simulation results are shown at 100 times their true size. The dotted, thin solid, dash-dotted and thin dashed lines represent fits to the MOA-II Sumi et al. (2013) all-source sample, MOA-II Sumi et al. (2013) RCG sample, MOA-II Sumi and Penny (2016) all-source sample, MOA-II Sumi and Penny (2016) RCG sample, respectively. The thick solid and dashed lines are fits to the resolved source and DIA source simulations of this work. The shaded areas represent the 68% confidence interval of MOA-II Sumi and Penny (2016) and Besançon fits, respectively

are lower than to Sumi et al. (2013) measurements. The revised maps are shown in Fig. 6.22.

Comparing the Sumi and Penny (2016) result to our Besançon simulation, the revised properties are still slightly higher (Fig. 6.23). The difference is in the error bar of the MOA-II measurement. Therefore, the revised MOA-II data are consistent the Besançon model *without any missing inner bulge population*.

6.7 Conclusion

A new version of the Besançon Galactic model is used to simulate microlensing optical depth, average timescales and microlensing event rate maps towards the Galactic bulge. The new model incorporates a refined two-component bulge (Robin et al. 2012). We perform a detailed comparison of the model with a recent optical depth study by MOA-II (Sumi et al. 2013) based on 474 events. The MOA-II observational filter, time scale cut and Gaussian kernel are applied to the maps. This is the first detailed field-by-field comparison between a theoretical microlensing model and a large-scale microlensing dataset.

In its original form the model overestimates the average time scale compared to the survey because the model lacks low-mass stars. Allowing for an extension of the model stellar mass function into the low mass star and brown dwarf regime, we find that the model correctly reproduces the observed average event timescale provided the mass function is essentially cut off at the hydrogen burning limit. The shape of the observed timescale distribution shows a weak evidence of an excess of short ($0.3 < t_E < 2$ days) and long ($30 < t_E < 200$ days) duration events and a deficit of moderate duration events ($2 < t_E < 30$ days). However, the model provides a satisfactory match to the MOA-II distribution (reduced $\chi^2 \simeq 2.2$).

Encouragingly, the inferred efficiency corrected MOA-II event rate is found to lie between the predicted number of events from the Besançon model for pure resolved sources and DIA sources. The number of Besançon microlensing events with resolved sources and DIA sources are 0.83 and 2.17 times number of MOA-II detected events. Given that the model analysis does not include a correction for blending in the number of available sources, and some expected differences due to differences in the assumed bandpass, this is in a reasonable level of agreement.

For the optical depth the residual maps between the model predictions and MOA-II observations show that there is generally good agreement over most of the MOA-II survey area and that the disagreement is confined to the regions closest to the Galactic Centre ($b < 3°$). The Besançon model predicts only 50% of the observed optical depth in this region. Maps of the event rate per star also show a similar disagreement. The fact that there is reasonable agreement in the maps of average duration but disagreement with the rate and optical depth argues for a mass deficit in the current bulge model or a systematic issue with the Sumi et al. (2013) analysis.

The bulge mass employed in the current Besançon model ($5.9 \times 10^9 M_\odot$) is somewhat lower than inferred in some recent studies such as Portail et al. (2015), which argued that a bar with a mass in the range $(1.25 - 1.6) \times 10^{10} \ M_\odot$ is compatible with recent radial velocity and proper motion studies. Such a massive bar could solve the optical depth discrepancy reported here. It remains unanswered whether such a model can be straightforwardly accommodated within a full population synthesis code. However, Robin et al. (2012) argued that the dust map model is likely to under-estimate extinction in the innermost regions due to incompleteness of 2MASS star counts below $K \simeq 12$. They also identify a missing population within the inner $\sim 1°$ in their model based on star count residuals. The additional population, along with increased extinction in this region should permit an increased optical depth without violating star count limits.

However, from recent a MOA-II study of Sumi and Penny (2016), the observed microlensing measurements were revised due to the incompleteness of stellar counts. The revised data are consistent with our Besançon results without any missing bulge population. Therefore, this validates MaBµlS use in predicting the microlensing properties of future microlensing surveys as well as consistency with large-scale datasets, such as, MOA-II.

References

Alard C (2000) A&AS 144:363
Alcock C, Allsman RA, Alves DR et al (2000) ApJ 541:734
Awiphan S, Kerins E, Robin AC (2016b) MNRAS 456:1666
Bond IA, Abe F, Dodd RJ et al (2001) MNRAS 327:868
Bond IA, Rattenbury NJ, Skuljan J et al (2002) MNRAS 333:71
Bramich DM (2008) MNRAS 386:L77
Hamadache C, Le Guillou L, Tisserand P et al (2006) A&A 454:185
Han C, Gould A (1995) ApJ 449:521
Holtzman JA, Watson AM, Baum WA et al (1998) AJ 115:1946
Hwang K-H, Han C, Choi J-Y, et al (2015) arXiv:1507.05361
Jaroszynski M (2002) Acta Astron 52:39
Jaroszynski M, Udalski A, Kubiak M et al (2004) Acta Astron 54:103
Jeffries RD (2012) In Reylé C, Charbonnel C, Schultheis M (eds) EAS publications series, vol 57. EAS publications series, pp 45–89
Kerins E, Robin AC, Marshall DJ (2009) MNRAS 396:1202
Kirkpatrick JD, Gelino CR, Cushing MC et al (2012) ApJ 753:156
Nataf DM, Gould A, Fouqué P et al (2013) ApJ 769:88
Penny MT, Kerins E, Rattenbury N et al (2013) MNRAS 434:2
Popowski P, Griest K, Thomas CL et al (2005) ApJ 631:879
Portail M, Wegg C, Gerhard O, Martinez-Valpuesta I (2015) MNRAS 448:713
Robin AC, Marshall DJ, Schultheis M, Reylé C (2012) A&A 538:A106
Robin AC, Reylé C, Fliri J, Czekaj M, Robert CP, Martins AMM (2014) A&A 569:A13
Sako T, Sekiguchi T, Sasaki M et al (2008) Exp Astron 22:51
Sumi T (2010) In: Coudé Du Foresto V, Gelino DM, Ribas I (eds)Pathways towards habitable planets, vol 430. Astronomical society of the pacific conference series, p 225
Sumi T, Penny MT (2016) ApJ 827:139
Sumi T, Abe F, Bond IA et al (2003) ApJ 591:204
Sumi T, Woźniak PR, Udalski A et al (2006) ApJ 636:240

Sumi T, Kamiya K, Bennett DP et al (2011) Nature 473:349
Sumi T, Bennett DP, Bond IA et al (2013) ApJ 778:150
Udalski A, Szymanski MK, Soszynski I, Poleski R (2008) Acta Astron 58:69
Wozniak PR (2000) Acta Astron 50:421
Wyrzykowski Ł, Rynkiewicz AE, Skowron J et al (2015) ApJS 216:12

Chapter 7
Summary and Future Works

In this thesis, wide-ranging studies on exoplanets and Galactic structure using the microlensing and transit techniques were presented. As the research in these areas is still progressing, there is a lot of things that can be done in the future.

Exomoons

If our Solar System is typical, then exomoons must be common. But, no exomoon has been detected. Only two exomoon candidates, MOA-2011-BLG-262L b and Kepler-1625 b I, are proposed (Bennett et al. 2014; Teachey et al. 2018). There are many programs that are trying to detect them using various methods. In Awiphan and Kerins (2013), we proposed another exomoon detection technique, using the correlation between TTV and TDV signal. In Chap. 4, we simulated the effects of intrinsic stellar variation of an M-dwarf host, which reduces the detectability correlation coefficient by 0.0–0.2 with 0.1 median reduction for *Kepler*-class photometry. For simulations with red noise with planet masses less than around 25 $M_\oplus$, 25–50% of them with 8–10 $M_\oplus$ moon have correlations high enough to confirm the presence of an exomoon.

Although, *Kepler* has ceased its main-mission operation, future space-based telescopes, such as the NASA TESS and PLATO missions, and ground-based observations should be able to detect exomoons in the near future. As the correlation technique requires long-term monitored fields, well-planned collected TTV and TDV data will help us discover exomoons via this technique. The observation should be done at the same wavelength, as the transit light curves have different shapes at different wavelengths, due to the limb-darkening effect of the host stars. Moreover, at present, we normally use modeling and fitting transiting exoplanet light curves programs, which do not include the transit of exomoons. Therefore, the measured TTV and TDV signals are the photocentric TTV and TDV signals, which do not provide sinusoidal signals as theoretical TTV and TDV signals.

© Springer International Publishing AG, part of Springer Nature 2018
S. Awiphan, *Exomoons to Galactic Structure*, Springer Theses,
https://doi.org/10.1007/978-3-319-90957-8_7

Transit Timing Variation

To date, more than 10 exoplanets have been detected using the TTV technique. In addition to detecting exoplanets, the TTV technique can also be used to detect exomoons. In Chap. 3, we presented the analysis of GJ3470b's TTV signal to search for the signal of a third body using the `TTVFaster` code, which computes the TTV signal from the analytic formulae of Agol and Deck (2016). Although, no significant TTV signal was found, the minimum mass of the perturber was estimated. For GJ3470b, the TTV signal shows little variation, which excludes the presence of another hot-Jupiter with orbital period less than 10 days in the system.

In the future, I plan to use the network of 0.5–0.7 metre robotic telescopes of NARIT, called the Thai Robotic Telescope Network, in Thailand, Chile, China, USA and Australia to continuously observe bright exoplanet systems, which can be used to detect the TTV signals and possibly discover additional exoplanets or exomoons via the TTV technique. This plan also includes a project for the selection of exoplanet targets for exoplanet atmosphere observations which I will collaborate with the University of Manchester, called SPEARNET (Spectroscopy and Photometry of Exoplanetary Atmopsheres Research NETwork).

Transmission Spectroscopy

At present, the transmission spectroscopy study is a highly successful technique for probing the exoplanet atmospheres, as it has been applied to several transiting exoplanets. In Chap. 3, the transmission spectroscopy analysis of the hot-Neptune, GJ3470b, which is the first sub-Jovian planet showing significant Rayleigh scattering slope, were presented. The result shows a low mean molecular weight atmosphere (1.08 ± 0.20) with atmosphere methane. A high altitude haze with dense tholin or polyaceylene is also suggested. However, the atmospheric models are single composition models. Therefore, a mix-ratio composition atmosphere model will be need to provide a better understanding of the atmosphere of the system in the future.

In the future, transit observations of GJ3470b at shorter wavelengths will aim to confirm the presence of the steep Rayleigh scattering slope and addition observations in the near-infrared will provide more detail of the GJ3470b atmosphere. As TESS and PLATO will be launched in 2018 and 2026, respectively, and will detect many bright nearby exoplanets. The number of transmission spectroscopy targets will rapidly increase. The study of them with small to medium sizes telescope, such as the Thai Robotic Telescope Network and the 2.4 metre Thai National Telescope, will provide a better understanding and better statistic of planetary atmospheres.

MaBμlS

Microlensing is the currently the most powerful technique to discover low-mass exoplanets beyond the snow line. We developed the Manchester-Besançon Microlensing Simulator (MaBμlS - http://www.mabuls.net) which is the first online real-time microlensing simulator, presented in Chap. 5. MaBμlS has capabilities to predict optical depth, average crossing-time and event rate for a specific area, filter and magnitude limit near the Galactic bulge. In Chap. 6, we performed a detailed field-by-field comparison between simulated MaBμlS data and 2006–2007 MOA-II observational data (Sumi et al. 2013), which is the first detailed field-by-field comparison between a theoretical microlensing model and a large-scale microlensing datase.

We modified the Besançon model to include M dwarfs and brown dwarfs. Our best-fitting model requires a brown dwarf mass function slope of -0.4. The model provides good agreement with the observed average duration, and respectable consistency with the shape of the timescale distribution. The model provides only $\sim$50% of the Sumi et al. (2013) measured optical depth and event rate per star at low Galactic latitude around the inner bulge ($|b| < 3°$). However, Sumi and Penny (2016) revised the star count of Sumi et al. (2013) data and found that the revised data are consistent without result without any missing bulge population.

In the future, we will try to improve MaBμlS to include an adjustable brown dwarf mass function slope for each population, because the real value of the slope is still ambiguous. The improvement will also include the addition of sources angular size and lens-source relative proper motion distributions, which can be used to provide real-time modelling of ongoing events. It will improve the observing efficiency of ongoing surveys and also precisely predict the microlensing properties of the future surveys, such as WFIRST and Euclid.

Moreover, due to a very small number of MOA-II microlensing events (474 events), we added the same brown dwarf mass function to all Galactic components which might not be a good representation. In the future, we will use the recently published microlensing events from the OGLE-IV survey (Mróz et al. 2017) to constrain the mass function of low brown dwarfs in the Besancon Galactic model. The data were obtained from 50 million stars within nine OGLE fields (12.6 square degree) in 2010–2015. The number of events in OGLE (2,617 events) is six times larger than the number of events in the MOA-II data (474 events). Moreover, the OGLE team published the histogram of the number of events in each Einstein crossing time range in each field, which can be used to estimate the mass function of brown dwarfs in each Galactic component.

References

Agol E, Deck K (2016) ApJ 818:177
Awiphan S, Kerins E (2013) MNRAS 432:2549
Bennett DP, Batista V, Bond IA et al (2014) ApJ 785:155
Mróz P, Udalski A, Skowron J et al (2017) Nature 548:183
Sumi T, Penny MT (2016) ApJ 827:139
Sumi T, Bennett DP, Bond IA et al (2013) ApJ 778:150
Teachey A, Kipping DM, Schmitt AR (2018) AJ 155:36

About the Author

The author was born in Thailand and graduated with a Bachelor of Science in Physics (First Class Honours) from Chiang Mai University, Thailand, in 2010. After finishing his undergraduate period, he was awarded a full scholarship from the Royal Thai Government to study in Master and Ph.D. degree in United Kingdom. He studied for a M.Sc. by Research (Astronomy and Astrophysics) degree at the University of Manchester from 2011–2012. He commenced studying for a Ph.D. (Astronomy and Astrophysics) with President's Doctoral Scholar (PDS) Award in September 2012 at the University of Manchester, the research aspects of which are presented in this thesis. He received his Ph.D. in Astronomy and Astrophysics from the School of Physics and Astronomy, the University of Manchester in 2016. At present, he is a researcher at National Astronomical Research Institute of Thailand (NARIT).

© Springer International Publishing AG, part of Springer Nature 2018
S. Awiphan, *Exomoons to Galactic Structure*, Springer Theses,
https://doi.org/10.1007/978-3-319-90957-8